D0878245

INVERTEBRATE BIOLOGY

Invertebrate Biology

A FUNCTIONAL APPROACH

P. CALOW

CROOM HELM LONDON

A HALSTED PRESS BOOK
JOHN WILEY & SONS
NEW YORK – TORONTO

Croom Helm Ltd, 2-10 St John's Road, London SW11

British Library Cataloguing in Publication Data

Calow, P.
Invertebrate biology.
1. Invertebrates
I. Title
592 QL362

ISBN 0-7099-0000-7
0-7099-0001-5 Pbk

Published in the U.S.A. and Canada
by Halsted Press, a Division of
John Wiley & Sons, Inc., New York

Library of Congress Cataloging in Publication Data

Calow, Peter.
Invertebrate biology.
"A Halsted Press book."
Bibliography: p. 166
Includes index.
1. Invertebrates. I. Title.
QL362.C28 592 81-6362 AACR2

ISBN 0-470-27238-4 Pbk

Typesetting by Elephant Productions, London SE15
Printed in Great Britain by offset lithography by
Billing & Sons Ltd, Guildford, London and Worcester

CONTENTS

Contents

ACKNOWLEDGEMENTS

I am grateful to the following holders of copyright for permission to produce various figures in the text:

Blackwell Scientific Publications for Figure 2.3 from *Symposium of the Royal Entomological Society* no. 6 (1973), Figure 4, p.19.

The Zoological Society of London for Figure 2.4 from *J. Zool., Lond., 193* (1981) Figure 1, p. 219.

Pergamon Press for Figure 2.6 from *Nutrition in the Lower Metazoa*, Smith, D.C. and Tiffon, Y. (eds) (1980) Figure 2, p. 19.

Blackwell Scientific Publications for Figure 2.8 from *Behavioural Ecology*, Krebs, J.R. and Davies, N.B. (eds) (1978) Figure 2.3d, p. 32.

Chapman & Hall for Figure 2.9 from *Ecological Stability*, Usher, M.B. and Williamson, M.H. (eds) (1974) Figure 1, p. 143.

Blackwell Scientific Publications for Figure 2.12 from *J. Anim. Ecol., 47* (1978) Figures 2a and b, p. 533.

Academic Press Inc. (London) Ltd for Figure 2.17 from *Advances of Ecological Research, 10* (1977), Figure 9, p. 35.

Annual Reviews Inc. for Figure 2.18 from *Annual Review of Entomology, 16* (1971), Figure 1, p. 374.

Elsevier/North-Holland Biomedical Press for Figure 3.7 from *Principles of Comparative Respiratory Physiology*, by Dejours, P. Figure 6.1, p. 71.

Blackwell Scientific Publications Ltd for Figures 6.7 and 6.8 from *J. Anim. Ecol., 48* (1979) Figures 1 and 2, p. 495.

Edward Arnold Ltd for Figure 6.9 from *Biology of Aphids* by Dixon, A.F.G. (1973) Figure 2.2, p. 9.

Cambridge University Press for Figure 7.4 from *Marine Mussels*, Bayne, B.L. (ed.) (1976) Figure 7.16, p. 282.

I am also grateful to Mr D.A. Read for compiling the indexes.
The jacket illustration of a scorpion feeding is redrawn by L.J. Calow from *Feeding, Digestion and Assimilation in Animals* by J.B. Jennings (Macmillan, 1972). Reproduced by permission of J.B. Jennings, Macmillan Publishers Ltd and St. Martin's Press, Inc.

PREFACE

Courses on the invertebrates have two principal aims: (1) to introduce students to the diversity of animal life and (2) to make them aware that organisms are marvellously integrated systems with evolutionary pasts and ecological presents. This text is concerned exclusively with the second aim and assumes that the reader will already know something about the diversity and classification of invertebrates. Concepts of whole-organism function, metabolism and adaptation form the core of the subject-matter and this is also considered in an ecological setting. Hence, the approach is multi-disciplinary, drawing from principles normally restricted to comparative morphology and physiology, ecology and evolutionary biology.

Invertebrate courses, as with all others in a science curriculum, also have another aim – to make students aware of the general methods of science. And these I take to be associated with the so-called hypothetico-deductive programme. Here, therefore, I make a conscious effort to formulate simple, some might say naive, hypotheses and to confront them with quantitative data from the real world. There are, for example, as many graphs in the book as illustrations of animals. My aim, though, has not been to test out the principles of Darwinism, but rather to sharpen our focus on physiological adaptations, given the assumption that Darwinism is approximately correct. Whether or not I succeed remains for the reader to decide.

As an aid to understanding the equations and graphs that are in the text and to identifying animals referred to by species the reader will find a glossary of symbols and a taxonomic index at the back of the book.

1 INTRODUCTION

1.1 Functional Biology – What Is It?

This book is about how invertebrate animals function – not just about *how* they work but also about *why* they work in the way they do. The term *function* means 'the work a system is designed to do', but in a biological context *design* is not quite the correct word, for organisms are not intelligently conceived nor are they intelligently selected. The characters we now see associated with organisms are the ones that, having arisen in the first place by chance, have persisted because they are better than others at promoting the survival of their bearers and their ability to reproduce. John Ray (1627-1705) and William Paley (1743-1805) thought that they saw evidence for the work of an intelligent designer in the organisation of living things but Charles Darwin (1809-82) replaced all that with a process based simply on mutation and a 'struggle for existence'. He called it *natural selection.* By *functional biology,* then, I mean the search for explanations of the success of particular traits in given ecological circumstances; or why, in other words, those traits which have turned up by chance have then been naturally selected. There is also a very important predictive side to the programme. What traits would be expected to evolve in particular ecological conditions?

In approaching this problem I have made a conscious effort, where possible, to apply the well-tried methods of science – formulating hypotheses and testing their predictions against data from the real world. Later in this chapter we focus on a number of general hypotheses concerning the way that natural selection is likely to have influenced the functioning of invertebrates, and these will be used as a basis for data-collection and data-evaluation throughout the rest of the book. The aims, however, are not to refute natural selection as a basis for explaining adaptation, but rather to use this hypothetico-deductive approach to sharpen our understanding of the functional biology of invertebrates. We assume, therefore, that evolution has occurred by natural selection and then attempt to discover what this means for how animals work.

1.2 The Invertebrates

'The line traditionally drawn in zoological teaching between the vertebrates (*animals with backbones*) and invertebrates (*animals without backbones*) is an unfortunate one; it obscures the fundamental unity that underlies the organization of living material' (*my italics*). Thus wrote E.J.W. Barrington in the preface to the first edition of his *Invertebrate Structure and Function* (Barrington, 1967). The line is, nevertheless, a convenient one since it separates a relatively small taxonomic group of animals with limited diversity (the vertebrates) in which a great deal is known about a few species from one of great diversity (the invertebrates) in which a little is known about many species. Since the adaptive *raison d'être* of structure and process is often made most clear by comparing species with widely differing life-styles and ecology, the diversity presented by the invertebrates is helpful for the functional approach.

A very brief classification of the invertebrates that will be discussed in this book is given in Table 1.1 and a scheme of possible relationships is given in Figure 1.1. For a more comprehensive treatment the reader must refer to more classical, zoological texts. Two good and complementary books are Russell-Hunter's *A Life of Invertebrates* (1979) and Fretter and Graham's *A Functional Anatomy of Invertebrates* (1976).

1.3 The Physiological Approach

Most of the processes seen in a phenotype involve the use of time and energy for personal survival and reproduction. The form of the phenotype depends on how resources from the food are allocated between different tissues and organs. Dynamic processes, such as metabolism and consequent growth and movement, depend similarly on a supply of resources. A study of the acquisition and allocation of resources by invertebrate organisms will form a framework for all that follows and the model illustrated in Figure 1.2 should be kept in mind throughout. This approach differs from more classical ones which either consider physiology on a group-by-group (Alexander, 1979) or system-by-system (Barrington, 1967) basis. Systems will certainly be considered here but only from the point of view of the way they are related to metabolic schemes of acquisition (Chapter 2), allocation (Chapters 3 to 6) and integration (Chapter 7). Control systems will be referred to throughout in so far as they are involved in the control of metabolism. However, the reader will not find detailed expositions on neurophysiology, sense

Table 1.1: A List of Major Invertebrate Phyla

Phylum	No. of species x 10^3	Major classes	Common name(s)
Porifera	4.2	Calcarea	Sponges
		Hexactinellida	
		Demospongia	
Ctenophora	0.08		Comb jellies
Cnidaria	11	Hydrozoa	Hydroids
		Scyphozoa	Jellyfishes
		Anthozoa	Anemones, corals
Platyhelminthes	15	Turbellaria	Flatworms/planarians
		Monogenea	Flukes
		Trematoda	
		Cestoda	Tapeworms
Nemertea (= Rhyncocoela)	0.6		Ribbon-worms
Rotifera	1.5		Wheel animalicules
Nemathelminthes (= Nematoda)	80		Round-worms, nematodes
Annelida	8.8	Polychaeta	Ragworms, lugworms etc.
		Oligochaeta	Earthworms etc.
		Hirudinea	Leeches
Mollusca	110	Monoplacophora	*Neopilina*
		Aplacophora	
		Polyplacophora	Chitons
		Gastropoda	Snails, slugs
		Bivalvia (Lamellibranchia)	Clams
		Scaphopoda	Tusk shells
		Cephalopoda	Octopuses, squids
Arthropoda	800	Onychophora	*Peripatus*/walking-worm
		Chilopoda	Centipedes
		Dipolpoda	Millipedes
		Insecta	Insects
		Crustacea	Crustaceans
		Merostomata	Horseshoe crabs
		Arachnida	Spiders, harvestmen etc.
Tardigrada	0.17		Water-bears
Echinodermata	6	Crinoidea	Sea-lilies
		Asteroidea	Starfish
		Ophiuroidea	Brittle stars
		Echinoidea	Sea-urchins
		Holothuroidea	Sea-cucumbers
Bryozoa	4		Moss animals
Brachiopoda	0.31		Lampshells
Hemichordata	9.1		

Figure 1.1: A phylogenetic tree giving a rough and very hypothetical summary of relationships between major invertebrate phyla. Broken lines delineate the sphere of interest for this book. Animals are not to scale.

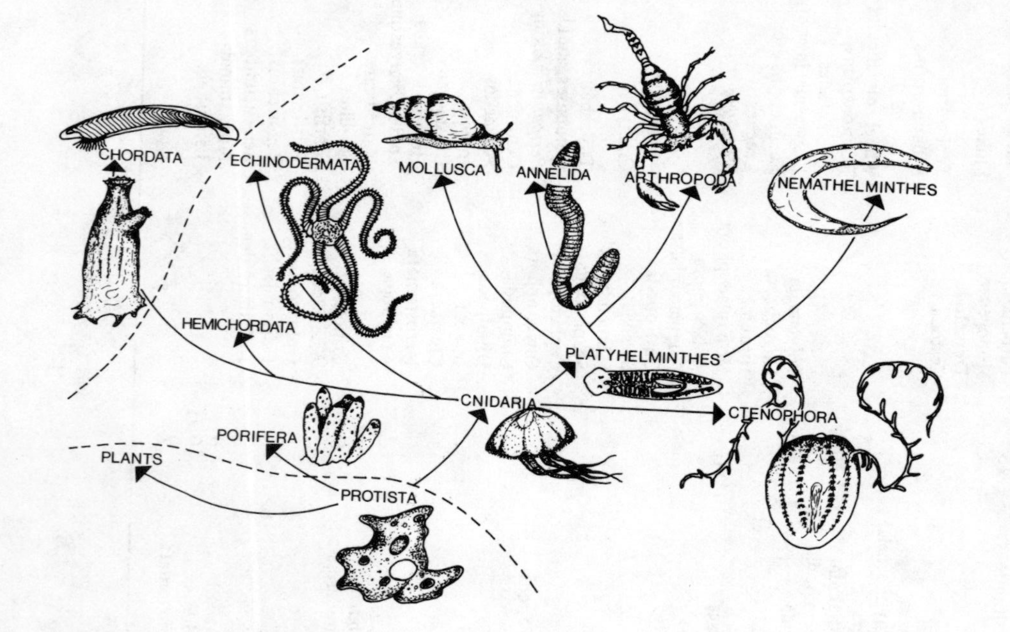

Illustrations by L.J. Calow.

Figure 1.2: Acquisition and allocation of resources in an invertebrate. Large box = whole animal; small boxes = sub-systems. I = ingested resources, F = faeces (egesta), I-F = absorbed resources (A); Rep = reproduction; Resp = respiration; H = heat; Ex = excretory products (excreta); E = environmental variables such as temperature, humidity, pH, Po_2, etc. I, F and A are treated in Chapter 2, Resp in Chapter 3, Ex in Chapter 4, Synthesis in Chapter 5, Rep in Chapter 6. The integration between sub-systems is considered in Chapter 7.

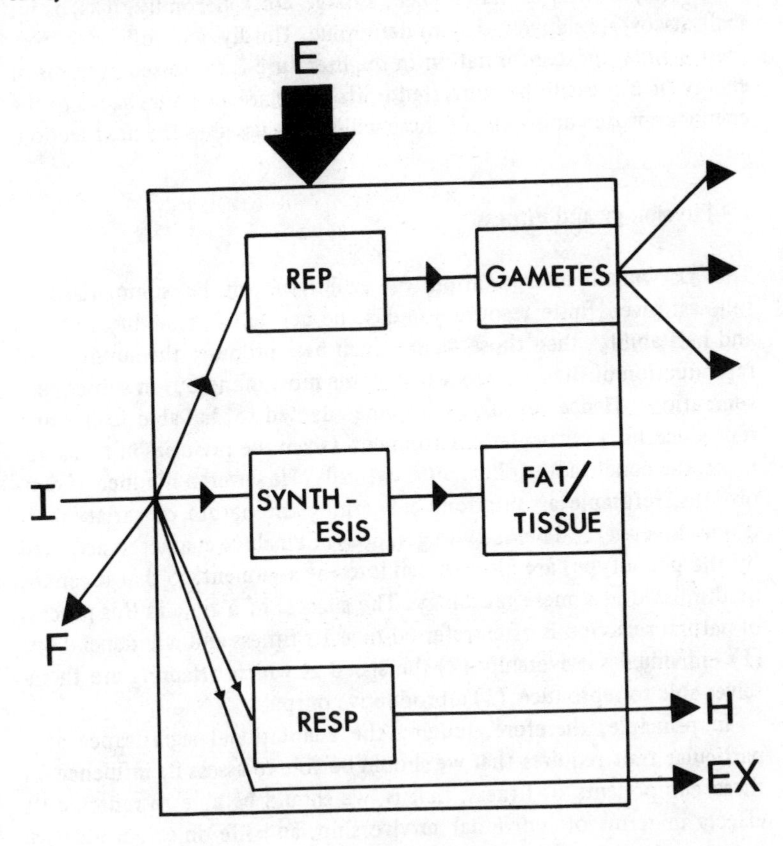

organs and endocrinology, and again explicit information on these must be sought from other sources (e.g. Barrington, 1967; Alexander, 1979).

The resources used by invertebrates are many and varied (see Chapter 2), involving a variety of organic and inorganic chemicals. The use of each in morphogenesis and metabolism could be expressed in terms of mole equivalents. However, since carbon occurs in all organic molecules and potential energy occurs in everything, measurements of either of these can be used in a general way to describe body size, reproductive investment and metabolic expenditure. In this book all components of an organism's resources will be considered, but special emphasis will be put on the energetic aspects. There are a number of reasons for this. First, and as already indicated, energy is a very general measure and can be used to quantify activities (feeding rates, respiratory rates, excretory rates, etc.) as well as states (size, storage etc.). Secondly, it is, as we shall discover, relatively easy to determine. Thirdly, and following from these points, most information in the literature is expressed in terms of energy or can easily be converted to it. There are also drawbacks to the energy approach and some of these will be discussed in the next section.

1.4 Physiology and Fitness

The Darwinian interpretation of evolution can be summarised as follows: given finite resources and some degree of variability in traits and heritability, then those traits which best promote the survival and reproduction of their bearers will become most numerous in subsequent generations. Hence organisms become adapted to, i.e. able to live and reproduce in, a particular environment. Given the premises in this argument, the conclusions follow automatically. However, a number of non-obvious, refutable assumptions concerning the nature of variation (it is non-directed) and heredity (e.g. it does not include characters acquired by the phenotype) are also written into the argument, so that it cannot be dismissed as a mere tautology. The success of a trait in this process of natural selection is often referred to as its fitness and will depend on: (1) individual survivorship; (2) the speed at which offspring are themselves able to reproduce; (3) reproductive output.

In principle, therefore, judging the adaptational significance of a particular trait requires that we should be able to assess its influence on these components of fitness; that is, we should be able to redefine its effects in terms of individual survivorship, and life-time reproductive success. However, it is usually very difficult, if not impossible, to translate short-term physiological processes (which may operate on a minute-by-minute basis) into their long-term demographic effects (which may operate on a year-by-year basis). More immediate if less direct measures of fitness are therefore required, preferably derived from the physiological processes themselves. One which has been used widely in this context is the 'net resource returns' from a particular process; i.e. the total gain credited to the process less the cost of effecting it. The rationale behind this is that the more resources that are available to the organism for building up its own tissues, the more rapidly it can, in principle, become reproductive, the bigger it might become as an adult and the more progeny it might ultimately leave. Other demographic factors being equal, all these characters are positively correlated with fitness as defined above and will be selectively favoured. Resources here might be measured in terms of carbon, nitrogen, essential amino acids or essential elements but as intimated above the *maximisation principle* is most commonly considered in terms of energy. As such it is often called the *energy maximisation principle.*

Of course, other demographic factors are rarely equal; for example, individual survival can be a costly business. The maintenance of biomass in a 'vital' and 'healthy' state is expensive in materials and energy, as are such processes as escaping from predators and catching prey. Furthermore, for reasons of constraint, all conceivable adaptations are not always feasible. The form and physiology of an evolving system may be compatible with certain kinds of change but not others – it seems unlikely for example, that the foot of a snail could ever evolve into a walking leg – and this kind of constraint is determined by prior evolutionary history. The processes of development, themselves, may allow only a sub-set of all conceivable transformations of genes to be expressed in a phenotype. Finally the genetic system, and the variation possible within it, may impose constraints on the phenotypes available for selection. Natural selection must, therefore, be thought of as a process which balances the benefits derived from metabolic costs against those derived from maximising the production of biomass, within a system of difficult-to-define constraints. This is often referred to as the *optimisation principle.* A consequence of both maximisation and optimisation principles is that minimum resources should be used in the building and maintenance of particular structures and in the support of particular processes, consonant with their effective operation, because more is then available for other aspects of the physiology of the animal. This is sometimes called the *economisation principle.* In what follows, we shall use all three principles, maximisation, economisation and optimisation in assessing the physiological adaptations of invertebrates. The general philosophy of this approach is discussed in Calow and Townsend (1981).

On the resources themselves, it is necessary to realise that selection is likely to have had the greatest impact on the utilisation of the ones which are most limited. Which resources are most limiting may vary from species to species and habitat to habitat and so our own emphasis should also change from one example to another. This is particularly true in the consideration of acquisition processes where animals may live surrounded by a sea of food rich in carbohydrate and energy but poor in nitrogen, as do many herbivores and detrivores. Alternatively, nitrogen and essential amino acids may be less of a problem for carnivores whereas energy may be very limiting. Since all feeding mechanisms are themselves limited in the amount of food they collect, however, these problems may be less important in determining allocation processes. Here all resources are limiting and, provided there is no interference between them, selection will have favoured 'wise' allocation of them all. Hence, for this reason, and for the sake of convenience, much emphasis in the book will be put on the allocation of potential energy as a resource.

1.5 The Last Word on Genetics

The above description of Darwinism is couched in phenotypic rather than genetic terms. We have paid only lip-service to the genes as being the ultimate controllers of phenotypic traits and this will be the case throughout this book. Darwin, himself, did not take much note of the hereditary mechanisms in his *Origin of Species* – principally because he did not know much about them – and Mendelian genetics was not grafted on to the theory until the 1920s. The marriage of evolutionary and genetic theory is referred to as Neo-Darwinism.

However, it has been argued that the phenotype is only an ephemeral unit – it is the genes which are enduring. This is because the genes collected into one phenotype are segregated, assorted and mixed with others by processes associated with sexual reproduction. Hence, in sexual species, when one phenotype dies its *unique collection* of genes goes with it. The appropriate, enduring unit of selection is therefore the gene, and evolution should be considered in terms of how genes spread through the gene pools of populations. If this is true, there are a number of points which are important for what follows: (1) it has to be assumed that all the physiological traits that we consider are determined genetically and that they are subject to the normal degree of variation; (2) it must be remembered that the way that genes behave (as summarised by the laws of genetics) may put constraints on the rates and even the course of evolution. On the other hand, the adaptational, phenotypic orientation casts certain doubts on the complete legitimacy of the 'selfish gene' idea because we discover that: (1) the phenotype is not wholly determined by its genes but by genes *plus* environment – i.e. development is said to be *epi*genetically not just genetically determined;

(2) the spread of a gene depends on the way the gene/environment-determined *phenotype* interacts with its environment – i.e. genes do not interact directly; (3) the phenotype works not as a mosaic of characters determined by dissociable genes, but as an integrated whole and the fitness of the genes it contains *depends upon* this integrated action. The critical reader must, therefore, keep these points in mind in evaluating all that follows. Nothing more explicit will be written about genetics.

2 ACQUISITION

2.1 Why Feed?

Invertebrates, like all other heterotrophic organisms, require food to supply the raw materials necessary for building the body and to supply energy to power movement and metabolism. Whatever its form or origin, food must therefore supply the following:

Proteins. These form much of the fabric of the body. They are needed for the growth, repair and replacement of tissues and are synthesised from amino acids derived from proteins taken up in the diet. Not all of the naturally-occurring amino acids involved in protein synthesis are required by heterotrophs since some may be synthesised from precursors. Those that are necessary, the precise requirements varying from one species to another, are called essential amino acids.

Carbohydrates. These are used mainly as fuel to power metabolism (chitin in the insect cuticle being one exception). For most invertebrates a great variety of hexose sugars are suitable since these molecules are freely interconvertible. Hexoses may be derived directly from food or from the enzymatic breakdown of polysaccharides like starch and glycogen.

Lipids. These include fats and several related compounds such as waxes, phospholipids and glycolipids. They are used mainly as long-term energy stores (Chapter 5), but are also important structural components of tissues, e.g. in cell membranes. Large lipid molecules in the food are usually broken down to emulsions and fatty acids before absorption.

Vitamins. These, like essential amino acids, are substances which heterotrophs are unable to synthesise for themselves but which are nevertheless required for normal growth, development and metabolism. They are often used as catalysts in metabolic processes and so are needed only in small amounts; in this respect they differ from essential amino acids. Vitamin requirements vary from species to species but less is known about this for the invertebrates than it is for the vertebrates. Arthropods appear to have much the same B-vitamin requirements as mammals, and a few insects, like locusts, also require vitamin C. For normal establishment and growth, the tapeworm, *Hymenolepis diminuta*, is dependent on some fat soluble vitamins being in the diet of its host but can do without vitamins A, D, and E and some vitamins of the B-complex. Another tapeworm, *Diphyllobothrium latum*, absorbs large quantities of vitamin B_{12} from the diet of its host and can, by so doing, cause anaemia. B_{12} must therefore be considered as a likely growth requirement for *D. latum*. Next to nothing is known about the vitamin requirements of most other invertebrates.

Energy. This measures ability to move matter or do work. The energy locked in chemical and particularly biochemical materials can be released as heat and to do work, but it need not be released at all. It is therefore known as potential energy. As with volume or mass, potential energy provides a quantitative measure of substance irrespective of whether it is ever used kinetically, and it is often employed in this way to measure the amount of food eaten by animals. This quantitative measure supplies little information about the quality of food in terms of the constituents listed above, though different biochemical substances do have different energy values (Table 2.1). In what follows, potential energy is usually expressed in joules (J) or kilojoules (1 kJ = 1000 J). The calories might also be found as a frequent measure of energy in the literature but the interconversion is straightforward; there are approximately 4.2 J/calorie and 4.2 kJ/kcal. Several methods are available for measuring the potential energies of biological material and one, micro-bomb calorimetry, which is commonly used for small invertebrates, is illustrated in Figure 2.1.

2.2 What is Eaten and How?

Food occurs in many forms. The supply is nevertheless finite and limited, so that selection will have tended to promote diversification in feeding habits and this should have resulted in the complete exploitation of all potential foods on this planet. Other things being equal, selection will, according to the maximisation and optimisation principles, also have favoured those animals which maximise the amount of food eaten for minimum costs and this will have led to considerable specialisation in feeding mechanisms and behaviour. In this section we review these specialisations, concentrating on morphological and overt behavioural ones. In the next section we turn to the more subtle aspects of behavioural adaptations. Three schemes of classification will be used to

Table 2.1: Energy equivalents of a variety of biological materials

	I kcal g^{-1} ash-free dry weight	II $kJ.g^{-1}$ (I x 4.2)
Aquatic algae	4.681	19.66
Terrestrial angiosperms	4.785	20.10
Porifera	6.475	27.20
Cnidaria	5.882	24.70
Platyhelminthes	6.332	26.59
Annelida	4.700-5.678	19.74-23.85
Mollusca (shell-free)	5.492	23.07
Arthropoda	5.445-5.673	22.87-23.83
Aquatic sediment	3.158-5.910	13.26-24.82
Terrestrial soil	2.275-3.412	9.56-14.33
Approximate values for biochemicals:		
Carbohydrates	4.1	17
Proteins	5.4	23
Lipids	9.0	38

Source: Cummins and Wuychek (1971).

give some order to the large quantity of information available on feeding. These are based, in turn, on the size of food particles, the trophic status of the food and finally on its mobility.

2.2.1 Particle Size

C.M. Yonge (1928) has divided animals into three major groups on the basis of the particle size of the foods they are adapted to exploit. Examples are illustrated in Figure 2.2.

(1) Microphagous Feeders. These are animals which ingest small particles (< 1 mm diameter) and are usually aquatic. They may feed on particles suspended in solution (suspension feeders) or deposited over and through the substratum (deposit feeders).

(a) *Suspension feeders* (Figure 2.2A) are highly diversified in taxonomic and ecological status. Their main requirement is a filtering device. This is formed from flagellae in sponges, cilia in many invertebrate phyla, setae which occur as fringes on various appendages of crustaceans

Figure 2.1: A micro-bomb calorimeter. A small sample of material, in the form of a pellet, is placed on the pan and a platinum fuse wire is brought into contact with it. The whole chamber is charged with oxygen, an electric current is passed through the fuse wire and the pellet undergoes complete combustion. An amount of heat is generated which is equivalent to the total energy potentially available from the sample. This is sensed by the thermocouples and is recorded on a potentiometer. The instrument is calibrated by burning measured amounts of material of known energy value (often benzoic acid) in the bomb prior to the determination of unknowns (see Phillipson, 1964).

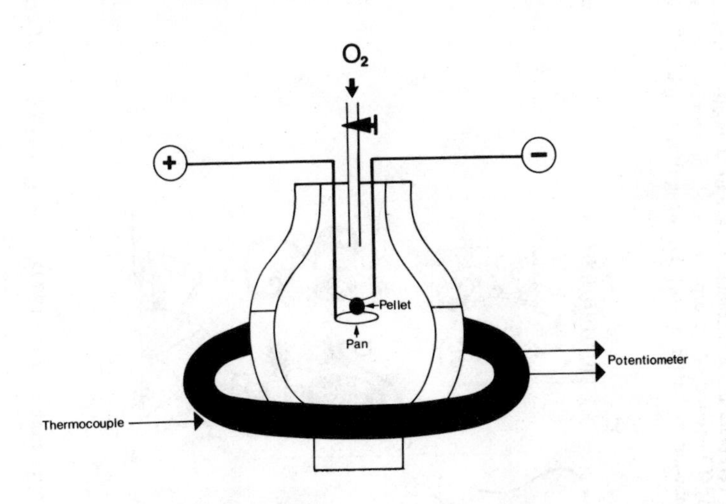

and some insect larvae, and tentacles which are used for filtration in sea-cucumbers. Mucous secretions play an important part in ciliary and tentacular feeding mechanisms as a final means of trapping food particles and in conveying them to the mouth. However, a few suspension feeders use mucus alone to trap food. For example, the sessile gastropod, *Vermetus gigas,* secretes a mucous veil from its pedal gland. This is used exclusively for the capture of food particles. The veil is used as a kind of net and, from time to time, it is hauled into the mouth of the snail and a new one is secreted.

(b) *Deposit feeders* (Figure 2.2B) are also very varied. Some use specialised mechanisms for collecting sediments; for example the mobile and ciliated tentacles of tubiculous polychaetes like *Terebella;* the mucoidal tentacles of sea-cucumbers and the specialised oral, tube feet of burrowing sea-urchins and brittle-stars. Other deposit feeders use unspecialised feeding mechanisms. Many burrowing polychaetes and oligochaetes, for example, use a simple mouth to eat their way through large quantities of sand, mud, silt and earth. The gastropod molluscs,

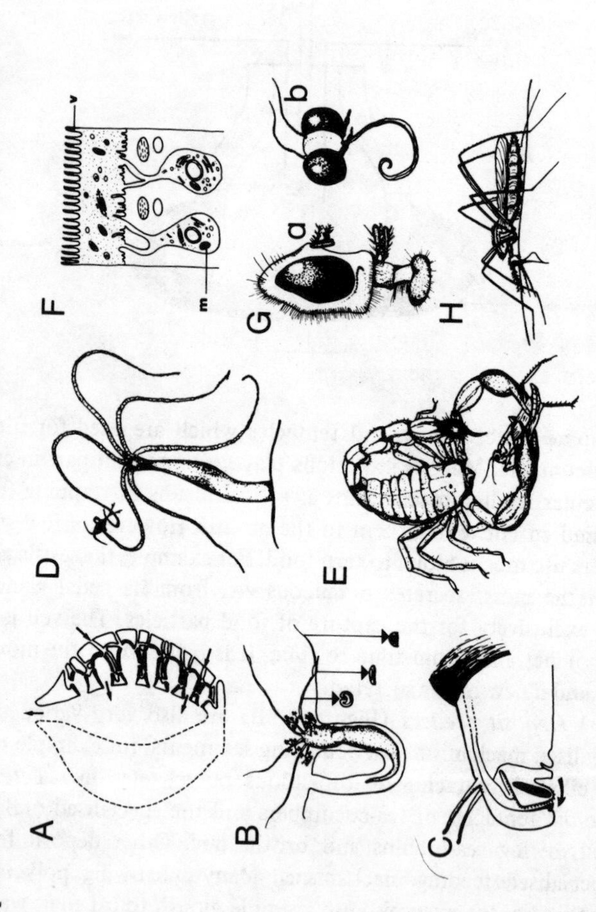

Figure 2.2: Feeding mechanisms classified on the basis of the particle size of the food processed. (A) Filtering chambers in a sponge – lined with ciliated choanocytes – arrows show water movement; (B) *Terebella* using tentacles to take up deposits – arrows point to cross-sections of the tentacle; (C) radula of a snail; (D) *Hydra;* (E) Scorpion; (F) section of absorptive surface of a tapeworm – v = microvilli; m = mitochondria; (G) sponging labellum of a fly (a), and suctorial proboscis of butterfly (b); (H) mosquito, piercing skin of host with sharpened mouth parts.

Illustration by L.J. Calow; Figure E redrawn with permission from Jennings, J.B. (1972). *Feeding, Digestion and Assimilation in Animals.* Macmillan: London.

intermediate in many respects between microphagous and macrophagous feeders, use a toothed ribbon, the radula, to scrape up small algae, detritus and bacteria (Figure 2.2C).

(2) Macrophagous Feeders. These can be divided into those animals which seize and swallow only and those which use some kind of pre-ingestion processing mechanism.

(a) *Seizers and swallowers* (Figure 2.2D) occur in almost every phylum. Cnidarians use tentacles, bearing a battery of stinging cells (the cnidoblasts) to seize prey and to deliver it to the mouth. Nemerteans capture prey by means of a long, slender proboscis which is sometimes barbed. Many polychaetes, like *Nereis* and *Glycera,* are carnivores and possess muscular pharynxes armed with jaws.

(b) *Seizers, masticators and swallowers* (Figure 2.2E) are more restricted in their taxonomic distribution. Cephalopods use a beak and a modified radula for shredding prey. Pincer-like chelae are found in some crustaceans (e.g. crabs) for seizing and shattering shell-fish. Mandibles are used for the mastication of both animal and vegetable material by insects, and modified parts of chelicerae and pedipalps are used for this purpose by those arachnids which masticate food.

(3) Fluid Feeders. These use nutrients which are dissolved in aqueous solutions and take them either by soaking, sucking or piercing and sucking.

(a) *Soaking* (Figure 2.2F) is used widely by endoparasites, and cestodes are good examples. These lack any form of mouth or digestive system and absorb food from the gut of their host through a specially modified cuticle with microvilli (to increase the absorptive surface-area and absorb host enzymes). The cuticle is also rich in mitochondria which supply the energy needed for the active transport of nutrients across the body wall. Free-living organisms in aquatic habitats are also surrounded by a medium rich in dissolved organic matter (DOM) – as much as 5 mg/l in seawater. After more than 100 years of debate, however, it is still unclear how important DOM is as a food for aquatic invertebrates. Polychaetes inhabiting sediments rich in organic material (which seep into the surrounding water) may meet major parts of their energy requirements by direct absorption of DOM but most freshwater invertebrates do not seem to use DOM to any appreciable extent (Jørgensen, 1976).

(b) *Sucking* (Figure 2.2G) is used by endoparasitic trematodes which have a suctional pharynx for this purpose, and is also well-developed in moths and butterflies where the insect mouthparts have been modified into a long proboscis for the exploitation of nectar. Muscid flies also have modified mouthparts for sucking. A sponge-like mechanism carried at the tip of the mouthparts of these flies soaks up organic fluids from the surfaces of plants and animals.

(c) *Piercing and sucking* (Figure 2.2H) occurs in a variety of animal- and plant-eating invertebrates. Turbellarian Platyhelminthes use enzyme secretions to push a pharynx into prey whereas lice, fleas and aphids depend more on physical pressure and sharpened mouthparts for insertion.

2.2.2 Trophic Status

Another way of classifying foods is in terms of their own trophic status (i.e. into autotrophs which can manufacture organic foods from inorganic precursors and heterotrophs which cannot) and their condition before being eaten (living or dead). Thus it is possible to construct the following scheme of classification of feeders:

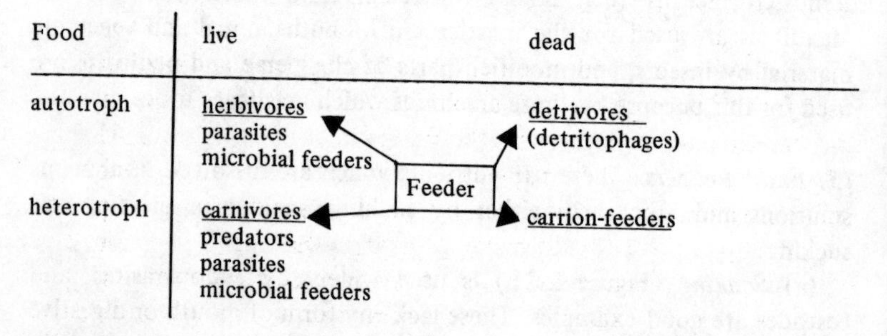

The distribution of some of these feeding categories (those underlined) through the major invertebrate phyla is summarised in Table 2.2, where an attempt is made to link this system of classification with that discussed in the last section.

Carnivorous feeding occurs in every invertebrate phylum but herbivorous feeding, particularly macro-herbivory, is more restricted. Plants have posed two kinds of problems for feeders. First, most large terrestrial plants (macrophytes) have evolved tough carbohydrate supporting structures, particularly cellulose and lignin, in response to life in a non-supporting atmosphere and, secondly, they have evolved specific defences against attack by the herbivores themselves. The latter include physical structures like hairs and spines, and chemical defences like alkaloids, glycosides, tannins, flavenoids, saponins and organic acids. The upshot has been that specialised and complex mouthparts, typical only of arthropods, molluscs and echinoderms, are required for ingesting

Table 2.2: Feeding modes classified on the basis of the trophic status and condition of foods. M = macro-particles; m = micro-particles; F =fluid.

	Fresh meat	Fresh plants	Dead materials
Porifera		m	m
Cnidaria	M		
Ctenophora	M		
Platyhelminthes	M/F	m	M
Nemertea	M/m		
Nemathelminthes	M/F	F	
Annelida	M/m	m	M/m
Mollusca	M/m	M/m	M/m
Arthropoda	M/m/F	M/m/F	M/m
Echinodermata	M/m	M/m	M/m

the tough vegetable structures, and biochemical evasion techniques are required for avoiding the action of the toxins. Furthermore, herbivores also have to come to terms with the indigestibility of cellulose and lignin, another possible reason for the evolution of these molecules, and with tannins, which, upon release from vacuoles within the tissues of plants, bind with proteins and also make them very resistant to digestion. The same hurdles are not encountered in the exploitation of soft meat so that it is not surprising that the carnivorous habit is more widespread than the herbivorous one (as suggested in Table 2.2).

The invertebrate group which has best risen to the nutritive challenge posed by plants is the Insecta. Indeed, in many respects these animals can be said to have co-evolved with the spermatophytes (i.e. the highest phylum in the plant kingdom). Even here, though, the number of taxa able to exploit plants is very restricted (Table 2.3).

Hard chitinised and specialised mouthparts have allowed insects to overcome the hurdle of ingesting tough vegetable materials. In response, however, plants have deployed not only non-specific toxins to ward off the attack, but also hormone-like steroids which are capable of inhibiting the growth and reproduction of the insects. The insects in their turn have evolved resistance. Indeed one possible explanation of why the number of species of insect associated with a species of plant increases in proportion to the abundance of that plant in recent geological history (Figure 2.3) is based on this response. As with insecticides, the

Table 2.3: Feeding habits of terrestrial insects (excluding blood-sucking and pollen-feeding species). * = major; + = minor.

	Scavengers and microbial feeders	Carnivorous	Fungi	Algae/lichens	Mosses/ferns	Spermatophyta
Symphyla	*				+	+
Pauropoda	*		+			
Protura	*					
Diplura	*	*				
Collembola	*	+	*	*		+
Thysanura	*					
Ephemeroptera	*	+				
Odonata		*				
Plecoptera	*	*		*		
Grylloblattodea	*					
Dictyoptera	*	*				
Orthoptera	*	*		*	+	*
Dermaptera	*	*				
Phasmida						*
Embioptera	*	*				
Zoraptera			*			
Isoptera	*					
Psocoptera	+		*	*		
Hemiptera		*	+	+	+	*
Thysanoptera	+	*	*			*
Mecoptera	*	*				
Neuroptera		*				
Trichoptera	*	*		+		+
Lepidoptera	+	+	+	+	+	*
Coleoptera	*	*	*	+	+	*
Hymenoptera	+	*	+		+	*
Diptera	*	*	*	+	+	*

Source: after Southwood (1973).

longer that toxins have been around, the more likely it is that insects will have evolved immunity to them. Once the defence has been breached, however, the plant has to invest more of its own resources in defence. Hence, there is a vicious circle driven, in a sense, by the maximisation principle itself, for the more effective is an evolved defence, the greater is likely to be the productivity of the defended plant and, because it is selectively advantageous for a feeder to maximise its rate of energy return from the food, the greater is the selective pressure on the feeder to breach the defence. One way out is for plants not to totally exclude feeders but to control how much of their tissues and what parts can be exploited, and it is possible that some of the chemical defences of plants are used in exactly that way (Maiorana, 1979). The 'sea' of green plants that we are confronted with in nature may not be the superabundant food supply that it appears. Control of feeding rates and patterns by chemical defences could explain why it is relatively undergrazed. Alternatively, there is much ecological information to suggest that the density of insect herbivores may be held below the potential carrying capacity of their foods by predatory control. Possibly both factors are important.

Figure 2.3: The relationship between the number of species of insect associated with a species of tree and the abundance of the tree in recent geological times. 'No. of records of Quaternary remains' is an index of the cumulative abundance of trees through recent history. The data are thoroughly discussed in Southwood (1961).

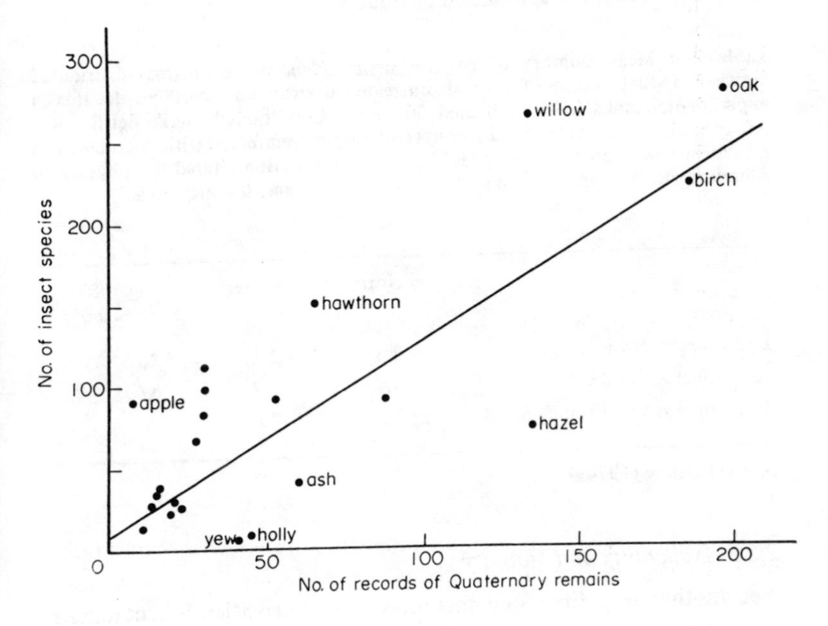

Source: Southwood (1973).

Resistance to plant toxins may take one of at least four forms: (1) reduced uptake across external surfaces (cuticle and gut); (2) increased excretion; (3) detoxification; (4) storage. All these mechanisms are employed by insects either singly or in combination. Some insects, notably beetles and butterflies, turn the plant toxins to their own advantage, either using them as olfactory identifiers or sequestering them in their own tissues for defence against predators.

Dead plant substances, which accumulate in soils and aquatic sediments, are even more difficult to exploit than living ones because they consist of accumulations of just those materials which are most difficult to degrade – cellulose, lignins, tannins and the like. They are poor in nitrogen and what little is present is usually in the form of tanned proteins. Detrivores therefore have low efficiencies of digestion (see Section 2.6) and compensate by moving food through their guts at a fast rate (Section 2.7). Some, possibly many, make more use of the microbes growing on the detritus than the detritus itself. Table 2.4, for example, shows how *Planorbis contortus*, a detrivorous, aquatic snail can, in laboratory feeding experiments, locate and preferentially feed on non-sterilised detritus. Some other detrivores, particularly soil arthropods, carry their own complement of detritus-attacking microbes around with them in specialised chambers in their guts.

Table 2.4: Mean number of aquatic snails (*Planorbis contortus*) distributed between a dual choice of normal, untreated detritus and sterilised detritus in experimental tanks (each replicated 30 times). Conditioned, sterile detritus was left in normal lake water for 24 hours and became reinfected with microbes. The results in row 3 preclude the possibility that sterilisation altered the physical or chemical nature of the food in such a way that it became less attractive.

	Normal	Other	Sig. diff. ($P < 0.05$)
1. Normal v. Normal	2.3	2.1	X
2. Normal v. Sterile	4.4	0.9	√
3. Normal v. Conditioned Sterile	2.5	2.8	X

Source: Calow (1974a).

2.2.3 *Availability and Mobility*

Yet another way that food materials can be classified is according to their availability and mobility. This influences the mobility of the feeders themselves and therefore provides another basis for classifying them.

Autotrophic foods are usually immobile, since locomotory powers are not required for trapping light (exceptions being many phytoplankton) whereas heterotrophic foods are more often mobile because this aids them in the capture of their own food and in the avoidance of being eaten. As a result of their distribution and abundance, some foods are readily available whereas others are scarce. This may alter seasonally.

The mobility and availability of the food will influence the effort that must be expended by the feeder in getting it. In general, natural selection should have acted to reduce the energy spent in food-getting because this will maximise the returns from each meal. The most economic strategy to deploy in the absence of food is to sit and wait for it to be replenished but, if in practice this is unlikely because the food is either scarce or immobile, then time and energy must be expended in seeking it out. Sit-and-wait feeders approximate to what Schoener (1971) has described as a TYPE 1 predator and the seek-out feeders approximate to what he describes as a TYPE 2 predator. This latter classification is widely used in the ecological literature.

Table 2.5 summarises what type of feeder is expected to exploit different kinds of food. Microphagous invertebrates, particularly filter-feeders, which exploit foods moved by water currents, adopt the ultimate in sit-and-wait strategies; i.e. they are permanently attached. Alternatively most herbivores, exploiting a food source which for most of the time is readily available, have to move to find food when it is in short supply because the food itself is immobile. For example, both the amount and quality of activity in *Ancylus fluviatilis*, a freshwater herbivorous gastropod which feeds on algae attached to submerged stones, are sensitive to starvation (Calow 1974b). In the presence of food this species slowly and systematically 'mows the algal lawn' so no particle is missed. At times when food is wanting, however, *A. Fluviatilis* becomes more active and its movement becomes more random as it switches to a searching mode. Top carnivores which exploit a precarious but mobile food, often adopt sit-and-wait strategies; spiders which sit and wait for prey beside webs are good examples. Alternatively carrion-feeders, or carnivores that exploit sessile prey, adopt seek-out strategies. Many freshwater planarians, for example, feed on active arthropods and adopt sit-and-wait strategies, whereas others exploit less mobile prey and adopt searching strategies. These differences are illustrated in Figure 2.4.

When food is in very short supply, extreme forms of strategies may be adopted. Animals may either enter a resting stage (undergo diapause –

Table 2.5: Distribution of trophic types relative to the availability and mobility of food.

Food Mobility	Food Availability		
	High	Low	Very low
High	*sit and wait feeders*		*migratory stages*
	microphagous	carnivores	*resting stages*
Low	*seek-out feeders*		
	heribivore detrivore	carrion- feeders	

period of arrested development, and aestivation – period of dormancy) or embark on extensive migration. Both strategies are usually adopted *in anticipation* of food shortage rather than as a direct result of poor feeding, since a food supply is necessary to lay down stores in preparation for the extended periods of inactivity or activity. Diapause occurs in anticipation of poor feeding conditions in Porifera (which form asexual, resting gemmules), *Hydra,* Cladocera, rotifers and many invertebrate parasites (which often form resistant sexual eggs). Aestivation is common in tropical and desert snails, usually prior to drought which is also a time of intense food shortage. Many marine, littoral invertebrates cease feeding and become quiescent at times of the year when food is in short-supply; this applies to some sea urchins, starfish and opisthobranchs. Mass, long-distance migrations are observed in several insect groups but notably in locusts and butterflies. These are not directly associated with food shortage but have undoubtedly evolved to prevent the consumers from overgrazing any one area.

Finally, it is important to note that which feeding strategy is adopted is also dependent on the physiology of the feeder, particularly in terms of its life-span and level of storage reserves. Thus in the Insecta, short-lived Diptera with low-level reserves cannot 'afford' to sit and wait for food and invariably become more active as they are starved (Barton Browne and Evans, 1960; Connolly, 1966) whereas long-lived cockroaches with extensive fat reserves, though exploiting dead and immobile foods, nevertheless become quiescent upon starvation (Reynierse *et al.*, 1972).

2.2.4 Energy Costs of Trapping Food

Some sit-and-wait predators use traps. Web-spinning spiders are an obvious example but there are others. Sit-and-wait planarians lay

Figure 2.4: Activity responses to starvation in several species of freshwater planarians. Probability of movement = proportion of those animals being observed which were active over the period of observation. One group (N = *Polycelis nigra*, T = *P. tenuis*, LU = *Dugesia lugubris*, P = *D. polychroa*) became more active over the first week that they were starved and these feed on immobile food. Another group (L = *Dendrocoelum lacteum*, B = *Bdellocephala punctata*) became less active over the same period of time and these feed on active arthropods. *Planaria torva* (TO) has an intermediate strategy and feeds on snails which are moderately active.

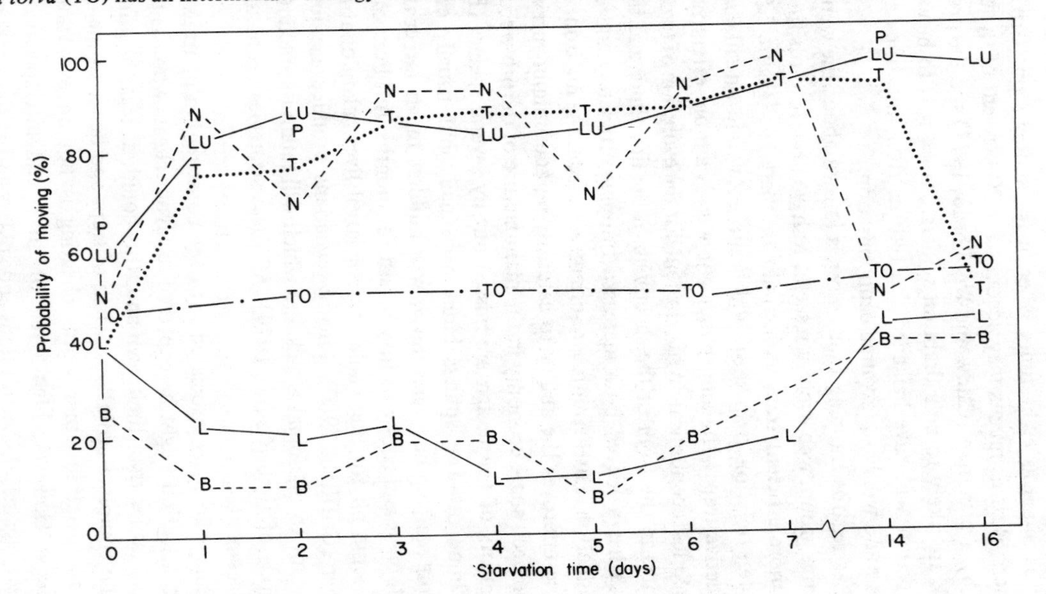

Source: With permission from Calow *et al.* (1981).

mucous patches which act as sticky traps and ant-lion larvae (Myrmeleontidae) dig pits and wait at the bottom with mandibles open, for prey to fall in.

The cost of building these traps is not insignificant. The metabolic rate of the linyphiid spider, *Lepthyphantes zimmermanni,* increases between two and three times over the resting level when it is building a web (Ford, 1977) and that of the ant-lion, *Morter obscurus,* increases by as much as eight times when it is building a pit (Griffiths, 1980). Energy lost as mucus may account for 20 per cent of the energy taken in by a planarian (Calow and Woollhead, 1977) and the proteins lost in silk are likely to be a drain on spiders. Hence it is to be expected that selection will have balanced the returns from the traps against these costs and will have favoured economisation.

As regards trapping food, most is known about the webs of spiders. In this group, web evolution seems to have been towards the production of more efficient traps with minimum materials (Figure 2.5). Primitive spiders rely on 'trip-lines' which are economic on silk but not very effective traps. The amorphous cob-web is a more effective trap but is much less economical on silk. In contrast the evolution of two-dimensional nets in the form of the sheet-web of the linyphiids and then the orb-webs has achieved both increased trapping efficiency and economy in production. Furthermore a re-orientation from the horizontal to vertical plane improves the ability to catch flying and jumping prey. There has even been some economy in the manufacture of the orb-web since whole segments of the design are missed out by several species (Figure 2.5).

Some orb-web spiders build webs daily but usually eat old ones before starting new ones. Sheet-web builders repair rather than replace their webs and cob-webs may persist for months. Ant-lions often occupy the same pit for the whole of their larval lives enlarging it as they get bigger (Griffiths, 1980). Finally, planarians produce a mucus which is resistant to microbial attack and which will retain its stickiness for long periods of time (Calow, 1979). All these strategies economise on trap replacement.

Since the costs associated with the construction of traps are significant, selection might be expected to favour cheats who use the traps of others rather than their own and, this could, at least in principle, lead to the complete evolutionary loss of traps and hence to the downfall of the trappers! In other words cheating is not an evolutionary-stable strategy. However, high levels of intra- and interspecific aggression militate against this in ant-lions and spiders and in the latter group there are surprisingly few species of klepto-parasites; i.e. species of spider

Figure 2.5: Spiders' webs. (A) trip-line; (B) cob-web – prey gets entangled in mesh and spider emerges from tunnel; (C) sheet-web – flying insects strike guy-lines and fall into sheet, spider moves in from beneath or side; (D) orb-web. Several orb-web builders economise by removing segments; (E) one to several omitted; (F) more omitted than produced; (G) central portion omitted.

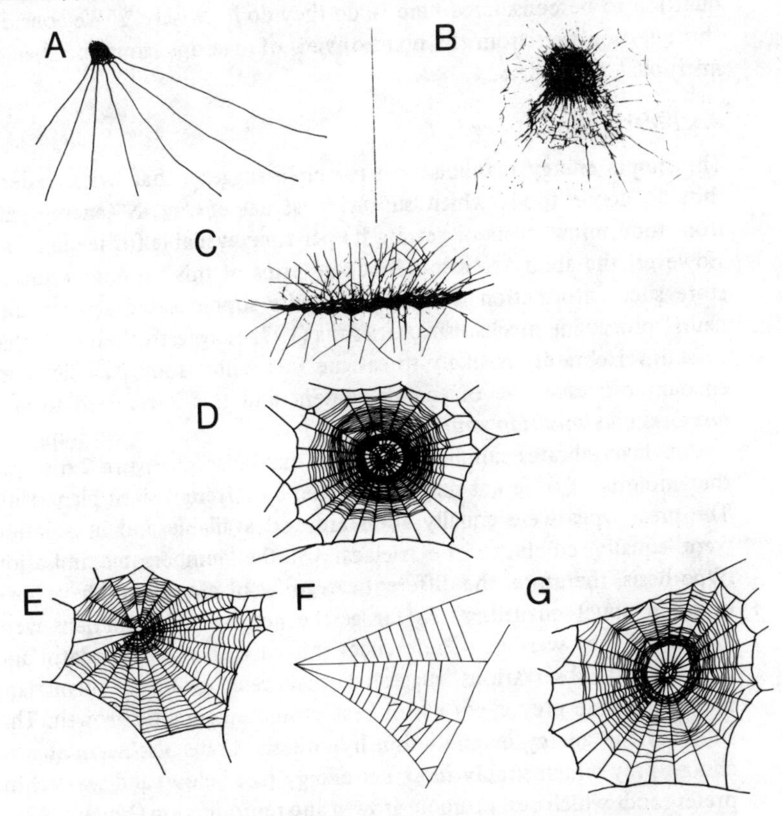

which specialise in using the webs of others. The factor which ensures that cheating does not do away with traps in planarians is that these animals require a mucous trail for locomotion. The trail lubricates the substratum and provides a point of leverage for cilia which give the motive force for locomotion in this group. Therefore, not even a cheating planarian could do without a mucous trail altogether!

2.3 Detailed Consideration of What Should Be Eaten

Within a particular trophic category some invertebrates eat a wide range of foods (they are said to be generalists or polyphagous), others eat a more restricted range (specialists or oligophagous) and some are committed to a single type of food (extreme specialists or monophagous). All animals exert some degree of choice over what they eat and the question to be considered here is 'do they do it "wisely"?' We consider this question first from the point of view of macrophagous and then of microphagous feeders.

2.3.1 Macrophagous Feeders

The simple energy maximisation principle suggests that 'wise feeders' should choose foods which supply most net energy, N (energy gain from food minus costs of getting it) per time available for feeding (T). However, the ability to assess foods in terms of this parameter and to store such information is likely to require sophisticated sensory and neural processing mechanisms. Griffiths (1975) suggests, therefore, that most invertebrates are likely to eat the first edible food particles they encounter irrespective of energy content and this is referred to as a *numbers maximisation strategy*.

Are invertebrates numbers or energy maximisers? Figure 2.6 shows the amounts of different-sized prey eaten by different-sized planarians. The prey types were equally abundant and available and in isolation were equally edible to the triclads. On the numbers maximisation hypothesis, therefore, the different categories of prey should have been eaten in equal quantities. As far as the non-starved planarians were concerned they were not. Preferences shifted with size of predator and independent observations suggested that each group of planarians preferred those prey types which best promoted somatic growth. This supports the energy maximisation hypothesis. Crabs also seem able to choose prey which supply most net energy (see below) and sea-urchins prefer foods which best promote growth and reproduction (Vadas, 1977). Clearly, at least some invertebrates can judge the value of food, but it seems likely that they need not be very 'clever' to do it – perhaps they simply use external features, such as size of prey, as a rough guide to suitability.

The total feeding time, T, can be thought of as being broken up into a series of feeding events: meal times t_m, separated by inter-feeding intervals, t_i. All other things being equal, net returns per T (i.e. N/T) will be inversely proportional to t_i because as t_i increases less meals are taken per T. t_i will itself be inversely proportional to the availability of food. Now consider Figure 2.7. This shows the consequence of including different numbers of food types in the diet for t_i and N/T. As more food

Figure 2.6: Small (S), medium (M) and large (L) planarians offered a choice of small (s), medium (m) and large (l) *Asellus.* ● – feeders starved for two days prior to the experiment; ■ – feeders starved for two weeks prior to the experiments.

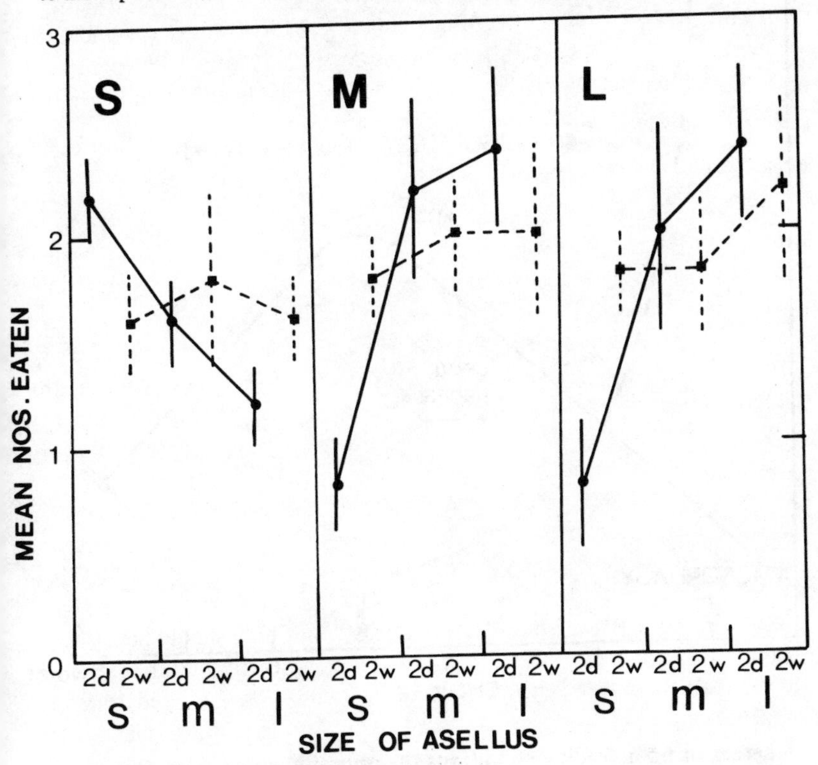

Source: With permission from Calow (1980).

items are added, t_i reduces, as does the average quality of food and its average energy content. N/T therefore rises to a peak as more and more food items are included in the diet (reductions in t_i are more important than reductions in food quality), but then declines (reductions in quality become more important than reductions in t_i). For an optimal diet all foods should be eaten to the peak. Clearly there will be a tendency to be monophagous or oligophagous when one or a few foods are very much more profitable than the rest (i.e. the peak will move to the left) and to be polyphagous as foods become less distinct in terms of the net energy they supply per feeding time (i.e. the peak will move to the right). In good feeding conditions, t_i will reduce and the peak will move to the left (i.e. feeding should become more specialised)

Figure 2.7: A graphical model of optimal diet. See text for further explanation.

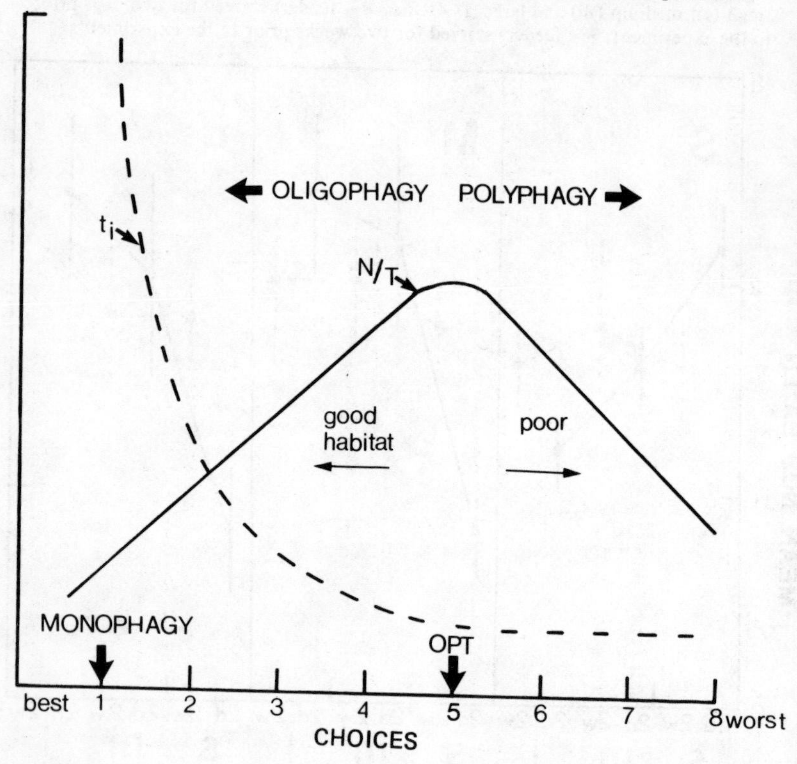

whereas in poor feeding conditions the opposite will happen. One piece of evidence from a field population of invertebrates which supports this prediction is that when food availability for the starfish *Leptasteria hexactis* is lowest (in summer) it eats 7-15 species of seaweed but when it is highest (in winter) it eats 3-9 species of seaweed (Menge, 1972). Another piece of supporting evidence, this time from the laboratory, is the starved planarians in the experimental situation described above also became less 'choosy' than satiated ones (Figure 2.6). Finally, Elner and Hughes (1978) carried out some laboratory experiments on the shore crab *Carcinus maenas* which hunts for mussels. They measured the energetic yield per unit handling and eating time of the crabs and found that big crabs get best returns from relatively big mussels. The results are illustrated in Figure 2.8. At high overall densities of food, the crabs totally ignored the smallest, least profitable size classes of prey whereas at low overall densities they were completely unselective.

Figure 2.8: Results from an experiment on the food choice of shore crabs eating mussels. Three prey sizes were offered and the largest were best in terms of N/T. The top row of histograms gives the relative availability of prey types. The second row gives the predicted optimal choice on the basis of a finite recognition time. The bottom row gives the results.

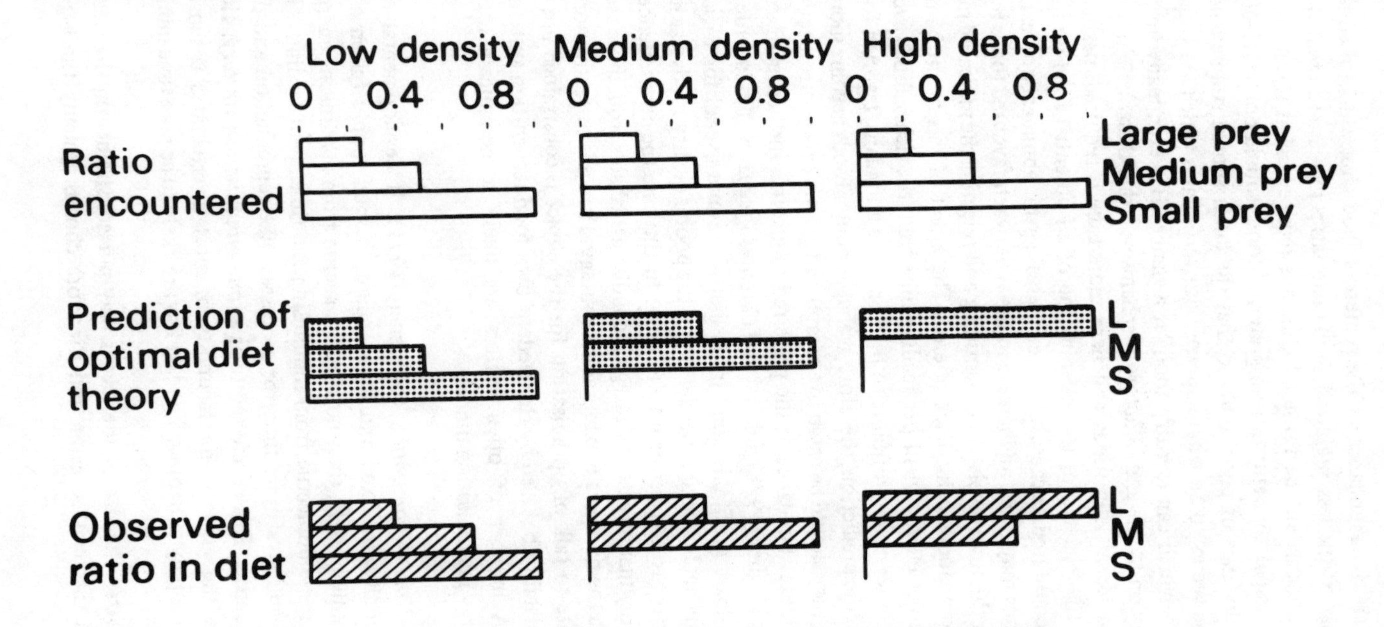

Source: The data are from Elner and Hughes (1978). The figure is taken with permission from Krebs (1978).

If the abundance of the preferred food choice is kept constant, but that of the less preferred food materials is increased so that the relative abundance of the favoured choice is reduced, what should happen? In principle, the returns from preferred foods should not be affected by the abundance of poor foods so that the latter should still be ignored, and this seems to be what happens with vertebrates which are visual feeders and which can identify food 'at a glance'. However, many invertebrates rely on tactile and olfactory mechanisms rather than on sight for locating and identifying foods and here more time is required, particularly by tactile-feeders, for the examination of *all* potential food items that are encountered. Because of this, an increase in the abundance of less favoured foods may cause an increase in the t_i between favoured foods by extending the time involved in examining and rejecting the increased number of poorer foods being encountered. The effect of this might be to reduce the net energy yield per feeding time from the richer foods such that it becomes more profitable to feed on more abundant, less-rich foods. The specific prediction, therefore, is that as poorer foods become more abundant they should be eaten to a greater and greater extent, irrespective of the absolute abundance of the preferred foods. In a further set of tests Elner and Hughes (1978) did find that *Carcinus maenas* would include more of the least profitable food in its diet if, despite overall food availability being held constant, the density of the poorer quality foods was increased relative to the better quality ones. In this case the consequences of the energy maximisation strategy are indistinguishable from those associated with the numbers maximisation strategy. Here it is certainly more appropriate to talk of optimisation for the choice is constrained not only by availability but also by the feeders' own food recognition system.

A number of other factors are likely to complicate the simple energy maximisation model:

(1) Time Minimisation (Schoener, 1971). If developmental rate and fecundity are fixed and independent of food supply (given a certain minimum level), then no benefits accrue from maximising energy gains but some do accrue from minimising the time spent feeding – because this will give more time for processess like reproduction and predator avoidance. This is referred to as a *time minimisation strategy*. However, since the growth and fecundity of most invertebrates is sensitive to food supply (Chapters 5 and 6) few are likely to be pure time minimisers.

(2) Size. The size of the feeder may put constraints on the size of the food that can be eaten and therefore chosen. Usually the bigger the

feeder the wider its choice in times of hardship. This has been demonstrated for copepods and ant-lions (Wilson, 1975).

(3) Nutrients Other Than Energy. Most herbivores and detrivores live surrounded by a sea of food but many still face limitations in nitrogen availability. Hence they may select food more on the basis of nitrogen than energy content. Fluid-feeding insects which exploit sap, like dipterans and aphids, can tap an unlimited supply of sugars but not nitrogen, and they do tend to choose foods more for their amino acid than their carbohydrate content (Dethier, 1976). The potential effects of toxins on palatability have already been mentioned.

(4) Switching. Many predators show frequency-dependent preference; that is, they take a disproportionate number of the commonest prey. This is known as switching and an example is given in Figure 2.9. Here, the predatory fresh-water bug, *Notonecta glauca,* was offered a choice between *Asellus* and mayflies and the abundance of the mayfly was increased relative to that of the *Asellus.* Under these conditions a larger proportion of mayflies was eaten at the high density than was expected on the basis of availability. The reason for this was that the attack success of *Notonecta* on mayfly improved as a result of experience; it learnt how to cope with the prey. The same was true if the abundance of the *Asellus* was increased relative to that of the mayflies. Hence, experience can alter preferences, by altering the costs and time associated with processing each prey. However, Murdoch (1969), working on predatory gastropods, has shown that switching will only occur if the preference difference between food choices is weak.

2.3.2 Microphagous Feeders

That microphagous feeders choose food on the basis of the energy maximisation principle is suggested by the following observation: deposit-feeding crustaceans, like hermit crabs and crayfish, sift sediment through setae and *reject the large particles.* Alternatively, suspension-feeding crustaceans, like copepods, use similar mechanisms but *reject the small particles.* This makes sense in energy maximisation terms since the energy content of algal cells increases with their size. Alternatively, if it is assumed that sediment feeders obtain most nutrient from the microbes covering the sediment rather than the sediment itself, then fine sediment is likely to give a better return than an equal quantity of coarse sediment because it has a larger surface area to volume ratio.

In that both mechanisms require the rejection of some particles, and

Figure 2.9: Switching in *Notonecta.* The straight line shows the result expected for no switching. *Asellus* were the alternative prey. o - are results at the start of experiment; ● – are results at the end of experiment. With one exception o did not deviate significantly from the 'no switch line', though the variation (represented as confidence limits) was large. However, experience caused switching.

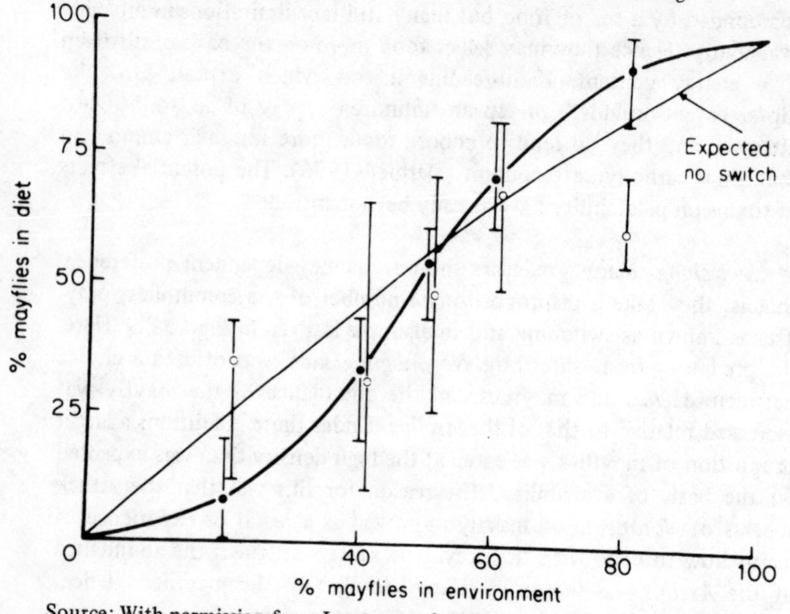

Source: With permission from Lawton *et al.* (1974).

in that this will incur a cost in terms of both time and energy, then the same principles may apply to microphagous feeders as apply to macrophagous feeders with finite recognition times. That is, as the density of the least-profitable particle increases, it is likely to become more profitable and should be eaten. Filter-feeding invertebrates do retain more small particles as their density is increased relative to larger, usually more preferred particles (e.g. see the work of Frost, 1977, on *Calanus*) but it is not known if this is due to active 'choice' or simply due to the effects of density changes in particle size-classes on the filtering device. The theory of feeding choice in these kinds of invertebrate is discussed at length in several papers: Lehman (1976), Taghon *et al.* (1978) and Doyle (1979).

2.4 How Much To Eat

There are two extreme types of eating strategy which can most easily be distinguished under conditions in which food is unlimited. These are illustrated as strategies A and D in Figure 2.10. In A, feeding is continuous, food passes continuously through the gut and the rate of input is equal to the rate of output. In D, feeding is discontinuous, the gut is filled periodically, food is processed, voided and the gut is refilled. Suspension feeders with a superabundant food supply would approximate to strategy A and a cnidarian or an asteroid echinoderm with unlimited food to strategy D. Of course in nature food is rarely unlimited and strategy A is more likely to tend to B or even C in Figure 2.10. Furthermore, inter-feeding intervals (t_i) are unlikely always to be completely dependent on the emptying of the gut since many feeders may begin to feed when the gut is only partially emptied if the opportunity arises. t_i also depends on food availability (see above). Hence strategy D is likely to approximate to C and even B in good feeding conditions.

Let us take strategies B and C as two distinct cases and consider what the maximisation principle means for each of them. Suspension and sediment feeders approximate to strategy B and most invertebrate predators to strategy C.

2.4.1 Strategy B

In principle, maximisation is best served in this strategy by maximising the amount eaten and hence the rate of passage of food through the gut. In practice two factors militate against this ideal: (1) feeding on or filtering small particles is metabolically expensive, requiring the constant action of cilia, flagellae or a radula, so that increased feeding rates will mean increased costs and these may increase at a disproportionate rate relative to feeding effort; (2) the more rapidly that food is passed through the gut the less well it is likely to be digested and absorbed (see Section 2.7). Figure 2.11(A) shows the compromise for a filter-feeder and is based on arguments in Lehman (1976). Concentrate first of all on the unbroken lines. The costs of filtering increase with filtering rate at an increasing rate but, because of the influence of rate of gut emptying on digestibility, the returns increase at a falling rate with filtering rate. The optimum rate is the one where the difference between the two lines is greatest since, here, net energy returns are greatest. If food concentration became reduced, the rate of return of food for a given filtering effort also becomes reduced (dotted line) and the optimum

Figure 2.10: Feeding patterns (m = amount eaten during a meal). (See text).

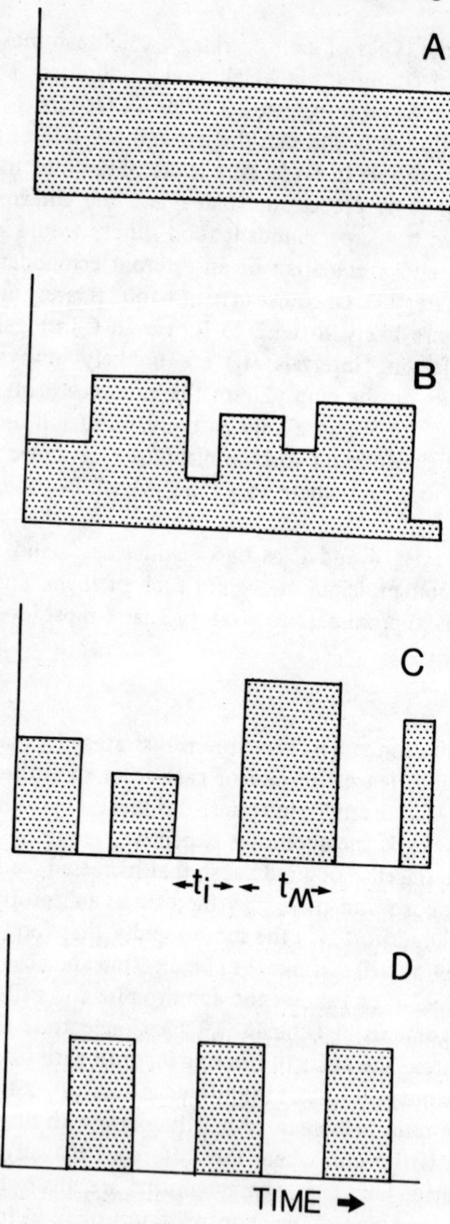

Figure 2.11: A – the model predicting optimum filtration rates. B – optimum filtration rates at different concentrations of food. Ingestion rate (= filtration rate x concentration of food) is also shown. C – actual filtration and ingestion rates for the crustacean, *Artemia* (redrawn from Reeve, 1963). See text for further explanation.

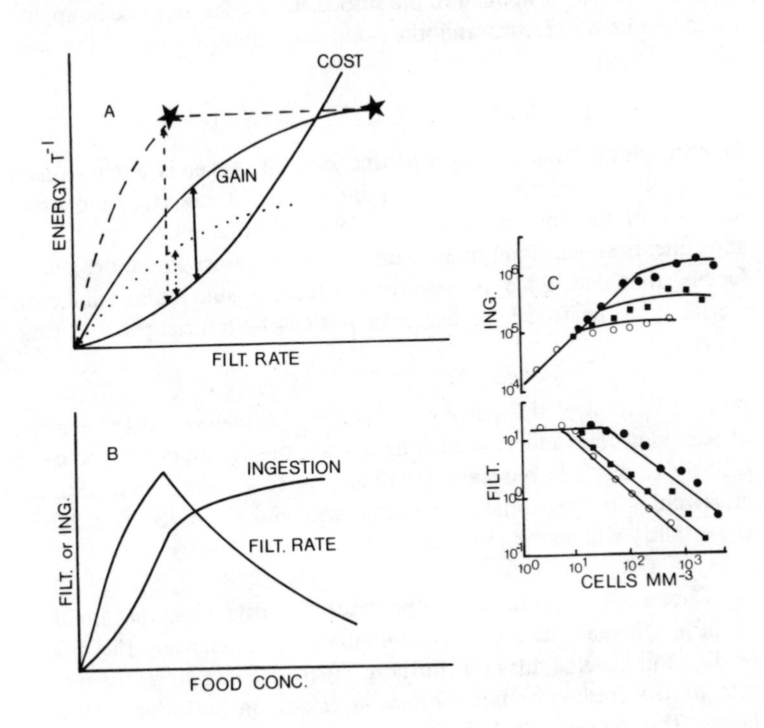

filtering rate lowers. The reverse should be true if the food concentration increased. However, the rate cannot increase without limit since the size of the gut is finite and the limit (★) will be reached at lower and lower filtration rates as the concentration of food increases (dashed line). Hence at some critical concentration, optimum filtration rate reduces with further increases in concentration.

A plot of optimum filtering rates against concentration is given in Figure 2.11(B). Ingestion rate (filtering rate x concentration) is also shown in this graph. An actual graph of ingestion and filtering rate against concentration is given in Figure 2.11(C) for *Artemia.* The former conforms exactly to expectation whereas the latter conforms to prediction over the high but not the low concentration and the discrepancy is common (Lehman, 1976). There may be several reasons for this. First, filtration rates are difficult to measure at extremely low concentrations of food, and secondly, water movements caused by filtering may also be required for respiratory purposes so that some will be necessary even at zero food concentrations.

The argument as applied to the suspension feeder may also be applied to the sediment feeder and the results are almost exactly the same (Doyle, 1979).

2.4.2 *Strategy C*

In principle, natural selection in this kind of feeder is likely to have resulted in the minimisation of t_i and of meal-times (t_m) and maximisation of the amount eaten per meal (m). This is because all these adjustments should tend to increase the total or gross amount eaten per feeding time and, since in these feeders feeding time is short and costs are likely to be trivial, to maximise net energy returns per unit time, i.e. N/T.

t_i. Minimisation of this parameter can only be achieved at the expense of searching costs and by increasing the rate of gut emptying. The latter is likely to result in reduced digestibility of food. When an investment in searching is appropriate has been considered above (Section 2.2.3). Digestibility will be treated later.

t_m. Once food is captured by a predator the costs of feeding are likely to be negligible. Hence t_m is minimised by maximising the rate of feeding and keeping this rate constant. Figure 2.12(a) shows the feeding rate of the freshwater bug *Notonecta glauca* on individual mosquito larvae. This predator feeds by sucking the body contents from its prey and the data in the figure were obtained by interrupting individual bugs after different lengths of time spent feeding. Contrary to expectation, *Notonecta* extracts food rapidly at first but the ingestion rate declines with time. This could either be because the feeder gets tired or because it diverts its attention from feeding to other priorities such as being more watchful for predators as it becomes satiated (McCleery, 1977). In this species, however, experimental interruption of feeding has shown that a constant rate can be maintained if feeders are continuously transferred to fresh food (Figure 2.12(b)). Hence the reduction in ingestion rate is a 'food' rather than 'feeder' effect and seems to occur because the last bit of food is more difficult to extract from the mosquito than the first. Exactly analogous results have been obtained for freshwater planarians (Calow, 1980).

Figure 2.12: (a) the cumulative dry weight of food extracted from mosquito larvae with time spent feeding by *Notonecta.* Prey size was (1) 4.0 (2) 3.6 (3) 3.0 (4) 2.8 (5) 2.0 mg wet weight. (b) The dry weight extracted when feeding for three minutes on successive, fresh larvae. The same conventions are used as in (a). The appropriate ingestion curves from (a) are also indicated by dots in (b).

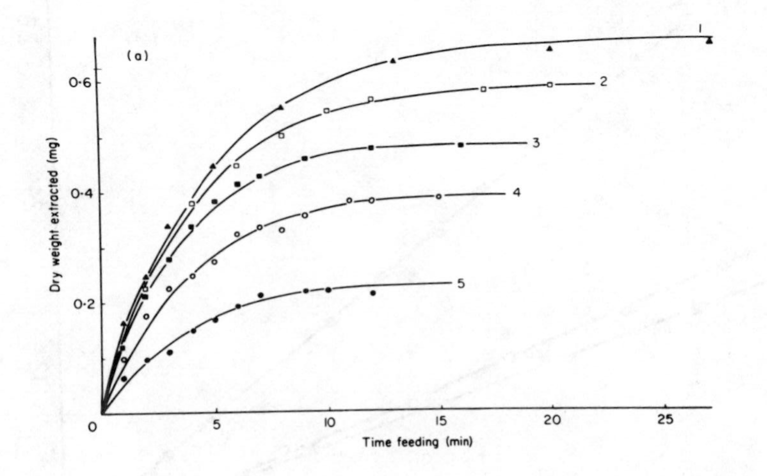

Figure 2.12 contd

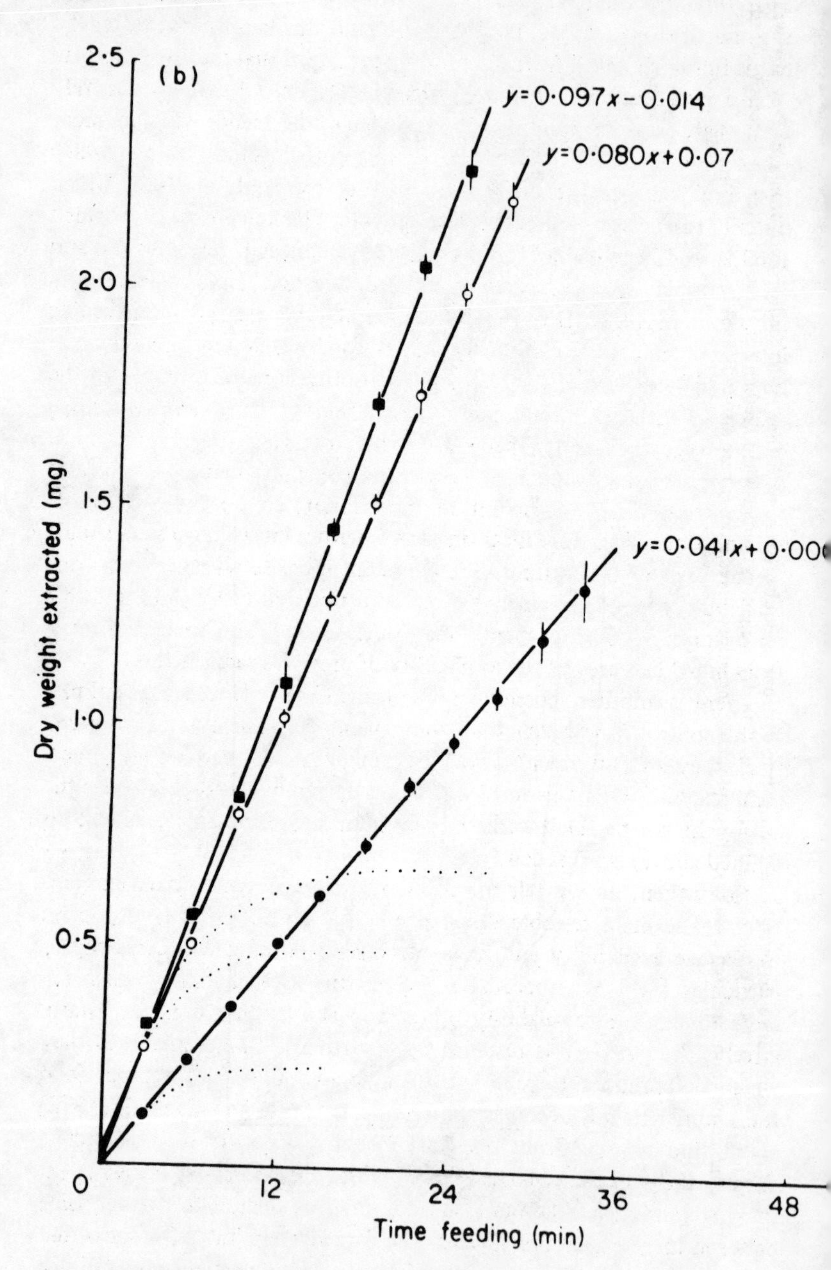

Source: With permission from Cook and Cockrell (1978).

The Amount Eaten Per Meal (m). This is determined in part by when the animal stops feeding, i.e. by t_m. Should this happen only after the gut is full or should it happen before then? Given that the rate of returns from a meal reduces as the meal progresses (Figure 2.12a) it is intuitively clear that, provided food is readily available, the feeder will gain more (maximise its overall returns) by leaving part-finished meals in which the food has become difficult to extract for fresh meals in which initially food can be extracted at a high rate. The key phrase is 'provided food is readily available' for as availability reduces it becomes less and less 'sensible' to leave parts of a meal uneaten. This is made more precise in Figure 2.13. Here $\hat{t}_i$ represents the average inter-feeding interval in a habitat with a particular food availability. The average intake ($\hat{m}$) per time ($\hat{t}_i + \hat{t}_m$) is given by the slope of a line from the origin to the appropriate point on the curve. The maximum slope occurs when the line is tangential to the curve (test it for yourself). If the shape of the feeding curve remains constant then as $\hat{t}_i$ increases (food becomes less abundant in the habitat) so does the optimum feeding time. Therefore meal times and, within limits, meal sizes should on the basis of the optimisation argument increase with reducing food availability as was anticipated. Cook and Cockrell (1978) did find that on average *Notonecta* spent more time feeding and extracted more from individual prey as the availability of prey was reduced.

There is another, possibly more straightforward, way that animals might control how much time they spend on a meal and how much food they extract from it, i.e. they might simply feed from a meal until they are full. This need not be an optimum response because the gut might not be filled until the optimum meal time and meal size (as defined above) is exceeded.

How, then, do we tell the difference between optimisers and gut-fillers? The main testable difference is that an optimiser is responsive to average availability and hence average $\hat{t}_i$ rather than necessarily to particular t_is. This is because it is only the average values which are indicators of the availability of food in the habitat at large. Alternatively, the gut-filler is responsive to particular t_is because it is they which determine gut space – the longer an animal goes without food the emptier its gut becomes. In *Notonecta,* Cook and Cockrell (1978) could find no significant relationship between particular t_ms and t_is despite the fact that, as has been noted above, the *average* time spent sucking (and hence amount sucked, $\hat{m}$) was related to *average* time between meals ($\hat{t}_i$) in this species. The data given in Table 2.6 show that the reverse is true for fresh-water planarians. Here particular t_is and t_ms

Figure 2.13: Model predicting optimal meal times (t_m) and meal sizes (m_o) for food at two levels of abundance; – = good, . . . = poor. (See text for details.)

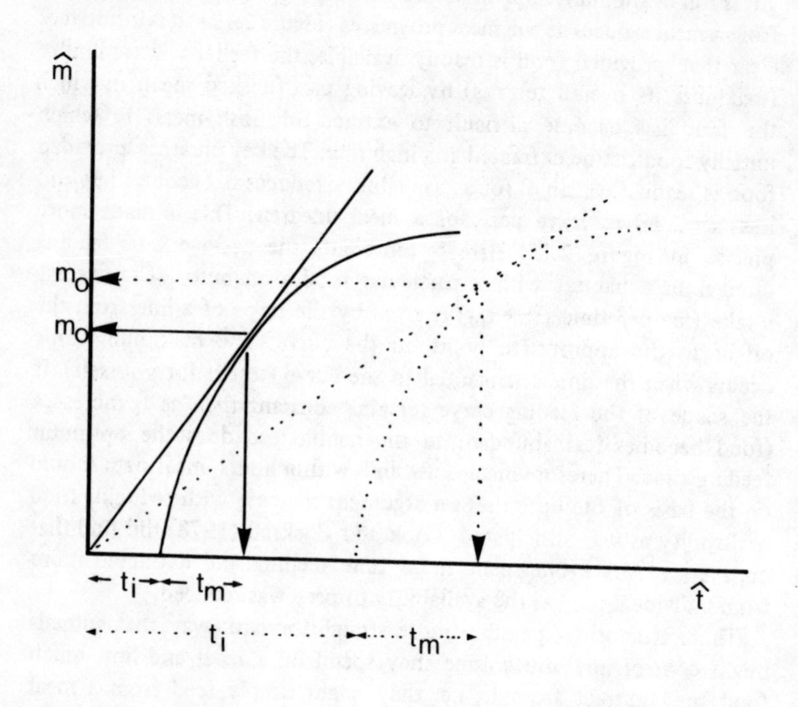

are more important than average ones in determing meal size. Hence *Notonecta* seems to be an optimiser whereas triclads seem to be gut-fillers. Of course gut-filling approximates to an optimisation strategy if t_i is always very large for in these circumstances the gut is likely to be filled before the optimum is reached, and this is likely to be the case in planarians which, as top carnivores, face regular shortages of food and must go for long periods without feeding. In gut-fillers, selection is likely to have favoured large guts to maximise intake when food is found and in the planarians we do observe much convolution and branching of the digestive chamber. Similar adaptations are found in leeches and blood-sucking insects. The constraints operating here are likely to be in terms of the metabolic costs of building, maintaining and carrying a large gut and these will be discussed further below.

Table 2.6: An experiment to test if a freshwater planarian, *Dugesia polychroa,* is a gut-filler or an optimiser (as defined in text). The data are mean dry weights of food extracted by predators from their prey (a freshwater isopod, *Asellus*). One group was fed once every two days, another once every six days and one on a randomly altering regime between two and six days. The last group was the crucial one, for here optimisers should respond to an average $\hat{t}_i$ (between two and six days) not to particular t_is, whereas the reverse is the case for the gut-filler. The results show that in this group the meals eaten after two days are always smaller than meals eaten after six days (and this was significant at $P < 0.05$). However, meals eaten after two and six days on the alternating regime were similar in size to the ones eaten on the corresponding continuous regime. Finally the first meals eaten after transference from one continuous regime to another immediately reflected the new t_i. *Hence planarians were responding to particular t_is not average ones and this suggests that they are gut-fillers.*

Feed	Once every two days (1/2)	Once every six days (1/6)	Alternating 1/2 and 1/6: After two-day intervals	Alternating 1/2 and 1/6: After six-day intervals
	0.66	1.48	0.47	1.66
	Transferred (1/6 → 1/2)	Transferred (1/2 → 1/6)		
	0.73	1.81		

Source: Calow (1980).

2.5 Gut Form and Function

In unicellular organisms digestion has to take place intracellularly after phagocytosis. This method of digestion is retained as a primitive feature in Porifera, Cnidaria, Ctenophora and the Platyhelminthes, occurs in some less primitive invertebrates which deliver finely-divided food to the gut (e.g. Rotifera, Tardigrada, Arachnida and the majority of Mollusca excluding Cephalopoda) and can be found, in some form or other, throughout the Invertebrata. The main advantage of intracellular digestion is that it is easier for animals to maintain an optimum concentration of digestive enzymes within a small, closed vacuole (hence economisation on protein synthesis). There are, however, a number of disadvantages viz: (1) intracellular digestion can only be associated with feeding mechanisms which deliver particles small enough to be phagocytosed; (2) every cell must be capable of secreting the whole spectrum of digestive enzymes needed to process each particle and this means that there can be no division of labour.

The beginnings of extracellular digestion and the associated elaborations of the gut occur in the Cnidaria and Platyhelminthes. Here some of the cells lining the sacculate and blind-ending digestive cavities become specialised as secretory cells and others as absorptive cells. An anus appears in phyla more advanced than the Platyhelminthes and the resulting one-way traffic of food material leads to considerable regional specialisation in the gut. In these invertebrates it is possible to identify at least five functionally distinct regions (Yonge, 1937). These are illustrated, in a schematic way, in Figure 2.14.

(1) Region of Reception. This includes the food-getting apparatus (discussed in Section 2.2.1) and the chambers into which food is received. Most invertebrates possess a buccal cavity – a simple chamber into which secretory glands empty. The latter may produce secretions which act as lubricants, have a proteolytic function (e.g. in some suctorial feeders), contain poison (e.g. some spiders), or act as anti-coagulants (e.g. blood-sucking leeches and insects). A few invertebrates, particularly the suctorial feeders, possess a muscular pharynx (or 'sucking stomach').

(2) Region of Conduction and Storage. This invariably consists of a conduction-tube, the oesophagus, but may also involve a storage region (crop). The latter forms the largest component of the gut of blood-sucking leeches and insects. Nectar collected by bees may be partially digested in their crops (by enzymes secreted in the buccal cavity) to form honey. The crop in the honey bee is often referred to as the 'honey stomach'.

(3) Region of Trituration (Mixing and Grinding) and Digestion. This often consists of a muscular region for grinding and mixing (the gizzard of gastropods and oligochaetes, the mastax of rotifers, the gastric mill of crustaceans and the proventriculus of insects) a region of ciliary sorting, and diverticulae which may be the site of digestion (sometimes intracellular) or simply the secretion of digestive enzymes. Digestive diverticulae occur as the caeca of polyclad Platyhelminthes, the hepato-pancreas of molluscs and crustaceans, the gastric caeca of insects and the pyloric caeca of asteroid echinoderms. In bivalves and some gastropods there is also a style sac. This contains a proteinaceous shaft (the style) which is rotated by cilia and pulls in a string of food-laden mucus, capstan-style, through the gut. The anterior end dissolves away and releases enzymes, mainly amylase, which have a digestive function.

Figure 2.14: A stylised and composite gut with buccal cavity (bc), salivary glands (sg), oesophagus (o), crop (c), region of trituration (t), region of ciliary sorting (cs), digestive diverticulae or caeca (dd), style (s) in a ciliated sac, intestine (i), malpighian tubule (m), rectal gland (rg) and rectum (r). No single invertebrate has all these structures. Shaded walls represent muscles.

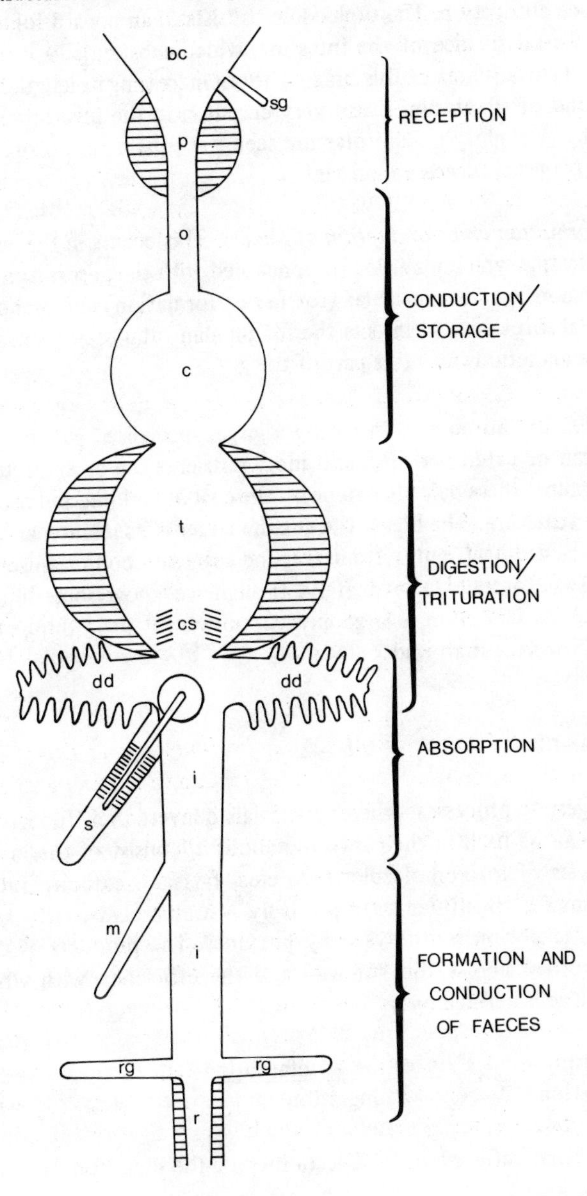

(4) Absorption. This often occurs largely, if not exclusively, in the digestive diverticulae. Otherwise the anterior portion of the intestine is modified for the absorption of food. Occasionally, specific structural modifications occur, like the typhlosole of terrestrial oligochaetes, to enhance absorption. The typhlosole consists of an inward folding of the inner dorsal surface of the intestine, which substantially increases the internal surface area of this organ without increasing its length. However, this kind of adaptation is not very common in the invertebrates and a straight or slightly coiled intestine seems adequate for absorption even when no gastric caeca are present.

(5) Formation and Conduction of Faeces. This occurs in the last part of the intestine which may also be concerned with the absorption of water. The region is often muscular (for faeces formation) and glandular (for water absorption). In insects the malpighian tubules also empty nitrogenous excretions into this part of the gut.

There are advantages in having a large, specialised gut in that more food can be eaten per meal and more nutrients can be extracted. However, against these gains has to be set the cost of building and maintaining such a structure. The bigger the gut the larger is its maintenance metabolism. Hence each gut system is to be seen as a compromise between these two forces. Unfortunately, though we know something of the gains to be had from a large gut we know next to nothing about the cost of carrying it around.

2.6 Digestibility

The digestive processes, whereby animals convert their food to a form which can be used in their own metabolism, consist of the progressive hydrolysis of macromolecules (proteins, fats and carbohydrates) into their simpler constituents (respectively – amino acids, fatty acids and sugars). Digestion is catalysed by enzymes. The products of digestion are absorbed across the gut wall and the efficiency with which this takes place is defined by:

$$\frac{\text{absorption}}{\text{ingestion}} \quad \text{or} \quad \frac{\text{ingestion} - \text{defaecation}}{\text{ingestion}}$$

and is often referred to as digestibility or the digestion or absorption efficiency (sometimes as the assimilation efficiency). Since the faeces often contain substances not derived directly from food, e.g. mucous bindings in molluscs and uric acid in insects, the second expression above often under-estimates the first. It is, therefore, sometimes referred to as apparent absorption efficiency or digestibility.

Enzymes are sensitive to temperature and most work best at an optimum temperature, falling off in activity above and below this level. Because of this, it has been suggested that absorption efficiences might be lower in 'cold-blooded' invertebrates than in 'warm-blooded' vertebrates and possibly may never exceed 30 per cent (Engelmann, 1966). However, maximum efficiencies recorded for most trophic groups do in fact exceed this limit by a wide margin (Table 2.7). Furthermore, for many invertebrates absorption efficiencies tend to be independent of external temperature over the normal range. These facts therefore suggest that: (1) the digestive enzymes of invertebrates are adapted to the temperature regimes in which they are required to operate and (2) several isoenzymes with different temperature optima may be involved in attacking each individual substrate.

Digestibility is likely to vary from one substrate to another and a broad generalisation which is widely quoted is that detritus (rich in the lignin and cellulose residues of plants) will be less easily and less well digested than live plant material and this in turn less well digested than meat. Table 2.7 indicates that soils and silts may indeed be intractable to digestion by invertebrates but plant materials (even those of macrophytes) seem to be digested as well as meat. Bacterivores, on which there are only few data, have high absorption efficiencies.

Protein production is expensive in energy and resources and so it is unlikely that all animals produce the same complete spectrum of digestive enzymes (economisation principle). Adaptive modifications in enzyme secretions are in fact common in the invertebrates (Mansour-Bek, 1954). The cnidarians and cephalopods, being carnivorous, concentrate on protease production as do many carnivorous gastropods. Herbivorous molluscs, however, produce amylases, glycosidases and cellulases instead. The same distinction also occurs between herbivorous and carnivorous arthropods and echinoderms.

More subtle distinctions may occur between related species which exploit slightly different foods. For example Calow and Calow (1975) found that the gut extracts from freshwater gastropods which feed preferentially on diatoms (with little cellulose) had lower cellulase activities than gut extracts of snails which feed on filamentous green algae (which are rich in cellulose).

Table 2.7: Absorption (assimilation) efficiencies in a variety of invertebrates.

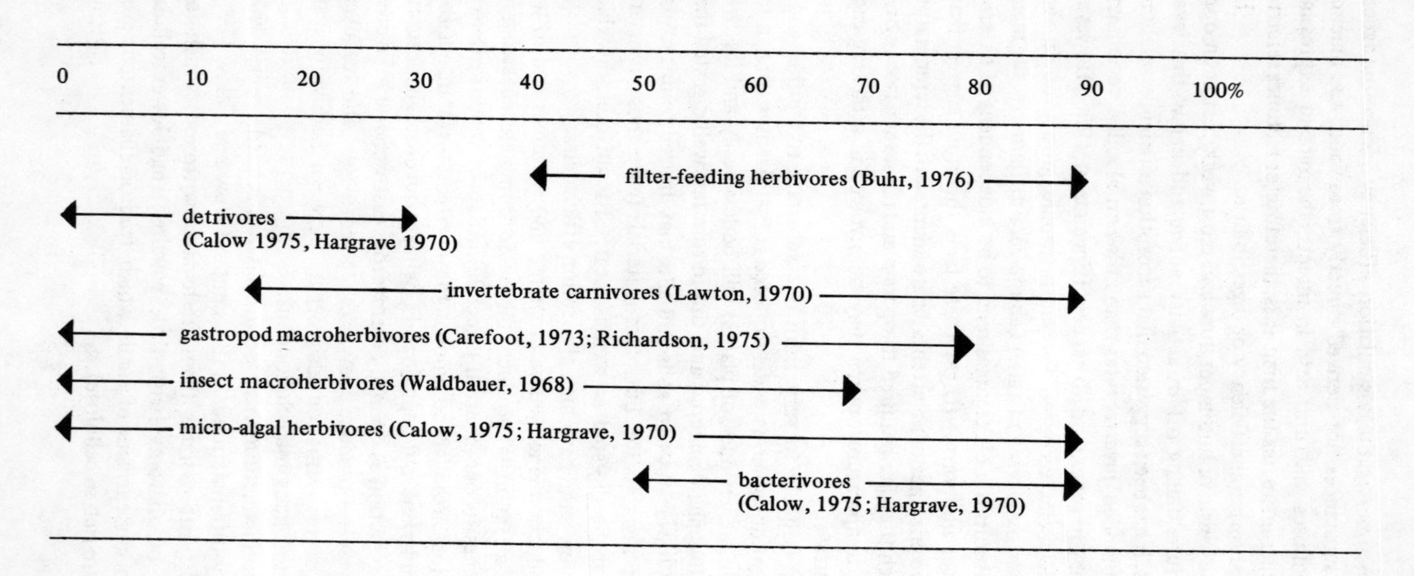

Intraspecific differences may also occur in enzyme secretion, particularly in species which metamorphose from one stage with one feeding mode to another stage with quite a different feeding mode. Lepidopteran caterpillars, for example, are adapted enzymatically for feeding on plants and produce proteases, carbohydrases and lipases. Alternatively the adults ingest only nectar and consequently secrete only carbohydrases and often only a sucrase.

Variations in enzyme secretion sometimes even occur in the same individual at different meals dependent on the nature of the food being eaten. For example, when larvae of the Black Carpet Beetle are fed, midgut proteolytic enzymes are secreted in proportion to the amount of ingested protein. Larvae fed a diet without protein show only a small increase in gut protease activity (Baker, 1978). Hence specific digestive enzymes seem to be secreted, probably on a quantitative basis, in response to specific classes of foods which are eaten. This process of induction is probably widespread though, as yet, there is little experimental information available on it.

2.7 Movement of Food through the Gut

A crucial determinant of digestibility and hence how much is obtained from a meal is likely to be the rate of passage of food through the gut. This is because the extent to which a meal is digested will depend upon how long the digestive enzymes have had to work on it.

Movement of food through the gut is brought about by ciliary mechanisms, muscles or a combination of both. Cilia are used by ciliary feeders (annelids and bivalves) though they may also be used by other invertebrates (rhabdocoel and polyclad Platyhelminthes and some gastropods). A combination of muscular and ciliary mechanisms occur in the echinoderms and many molluscs. Transport of food exclusively by a muscular system occurs in all Arthropoda. Many of the worm-like invertebrates, such as planarians, rely a great deal upon the contraction of the body wall musculature to move food from one part of the gut to another.

Unfortunately there are few data on the quantitative relationship between the amount absorbed from a meal and the time that it is retained in the gut, but it seems likely that the relationship illustrated in Figure 2.15 for an octopus, where the cumulative amount extracted increases, but at a reducing rate, with retention time, is a general one. Therefore, consider the general argument illustrated in Figure 2.16. The rate of obtaining resources is the amount obtained divided by the time

Figure 2.15: Left figure shows cumulative uptake of food from a meal undergoing digestion in *Octopus* gut (as a percentage of that digested). Right figure shows simultaneous, cumulative loss of faeces (per hr) as a percentage of that ingested.

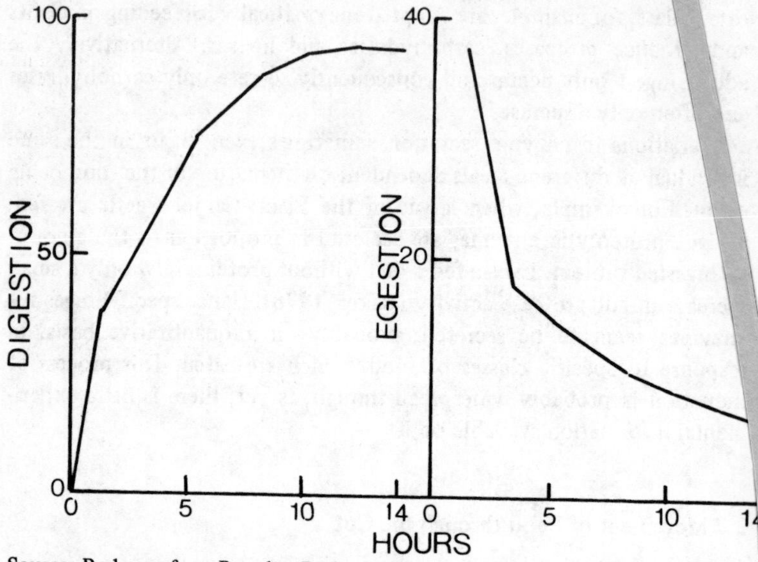

Source: Redrawn from Boucher-Rodoni (1973).

taken to obtain it, i.e. the slope of a line from the origin to any point on the curve. The rate of obtaining energy is maximised by the steepest line, and this is the one which just touches the cumulative uptake curve, i.e. is tangential to it (this theory is elaborated in Sibly, 1981). This is the optimum retention time. It is possible, however, that for some animals the cumulative uptake is a linear function of retention time and this is particularly likely for fluid feeders where enzymes can mix freely with the meal throughout. Here the optimum response is for the food to be retained for as long as is needed to digest it all.

Figure 2.16 suggests two predictions for the curvilinear case. These are that as the quality and availability of food reduce, retention time should increase. Conversely, very rapid throughputs of food are expected in continuous-feeders, since the inter-meal interval is effectively zero. Here only the costs of obtaining food at a high rate and of moving it rapidly through the gut are likely to militate against rapid rates of gut-emptying.

The few data available on gut emptying in invertebrates tend to confirm these ideas. Continuous-feeding sediment-eaters have fast rates of throughput (e.g. the earthworm *Allolobophora rosea* may completely

Figure 2.16: Cumulative absorption of materials and/or energy from a meal. There is an initial lag phase representing the time required for the digestive enzymes to get to work. Uptake rates (A time $^{-1}$) at any point on the curve are obtained by drawing straight lines from the origin to the points in question (e.g. ab or ac). The fastest rate is found when one of the lines is tangential to the curve (i.e. ac). Therefore ac defines the optimum retention time. With a poor quality food the lag phase is increased, the rise of the curve becomes less steep and the maximum possible returns from the meal reduced. ac would therefore reduce the slope and predict a longer optimum retention time. The lag phase is also influenced by the rate of encounter with meals, lengthening as food availability reduces. Hence, other things being equal, because the optimum retention time shifts to the right as lag phase increases then food should be retained for longer in poor feeding circumstances. These arguments are similar to those given in Figure 2.13 and are from Sibly (1981).

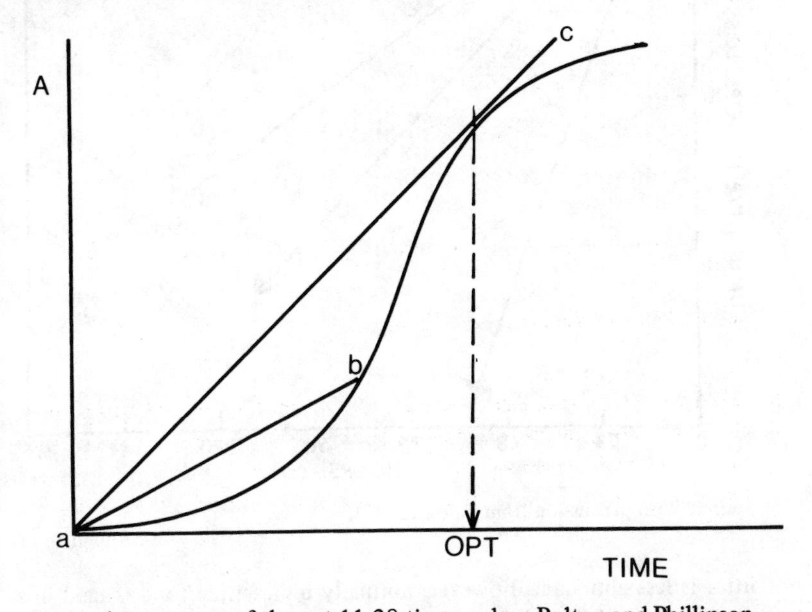

change the contents of the gut 11-28 times a day: Bolton and Phillipson, 1976) whereas blood-sucking insects and leeches, in which meals may be few and far between, retain food for weeks and even months. Figure 2.17 shows the effect of a reduction in the availability of food on the rate of passage of a non-absorbed isotope, ^{51}Cr, through the gut of the freshwater snail *Ancylus fluviatilis.* As expected from the theory a period of starvation before the administration of the isotope resulted in a reduced rate of throughput. The same was true if snails were starved after being fed on the labelled food. Similar phenomena have been reported for terrestrial isopods (Hassall and Jennings, 1975). Here, when litter is abundant, feeding is continuous and the gut is full. When

Figure 2.17: Defaecation strategies in a freshwater limpet measured by cumulative loss of a ^{51}Cr-labelled meal. a = snails fed before and after labelling; b = starved before but not after; c = starved after but not before.

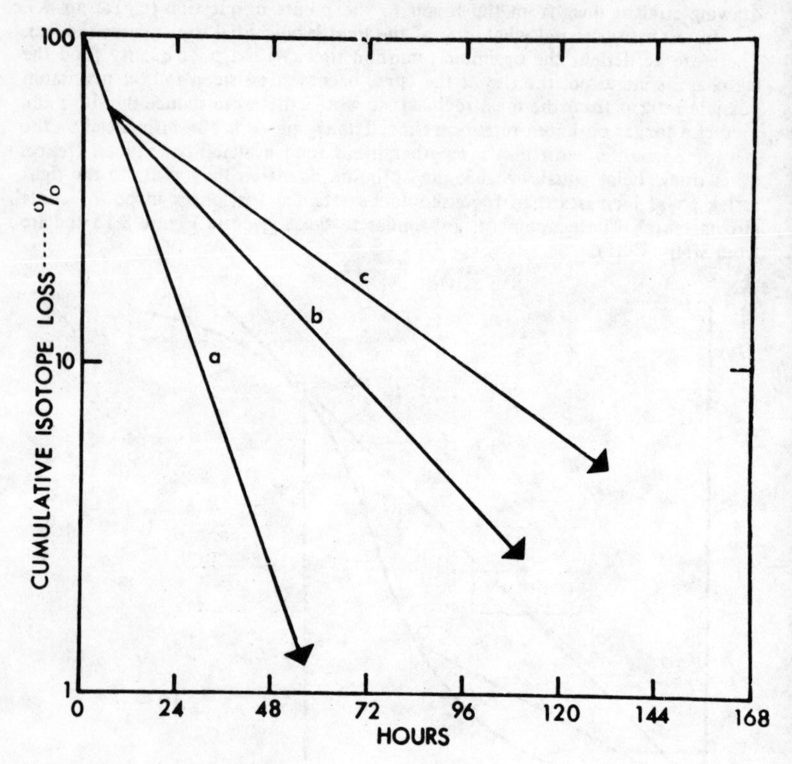

Source: With permission from Calow (1977).

litter is less abundant, however, normally indigestible food is held in a distinct part of the gut for an extended process of digestion effected by cellulases from micro-organisms ingested with the food. In both snails and isopods these 'holding tactics' lead to greater absorption efficiencies.

2.8 Control

The regulation of what foods should be eaten, how much should be ingested and at what rate, and the rate of digestion, absorption and defaecation of a meal implies the existence of sophisticated control mechanisms. This in turn suggests the existence of sensors and feed-back

Figure 2.18: Mechanism for metabolic homeostasis in the blow-fly (see text for further discussion).

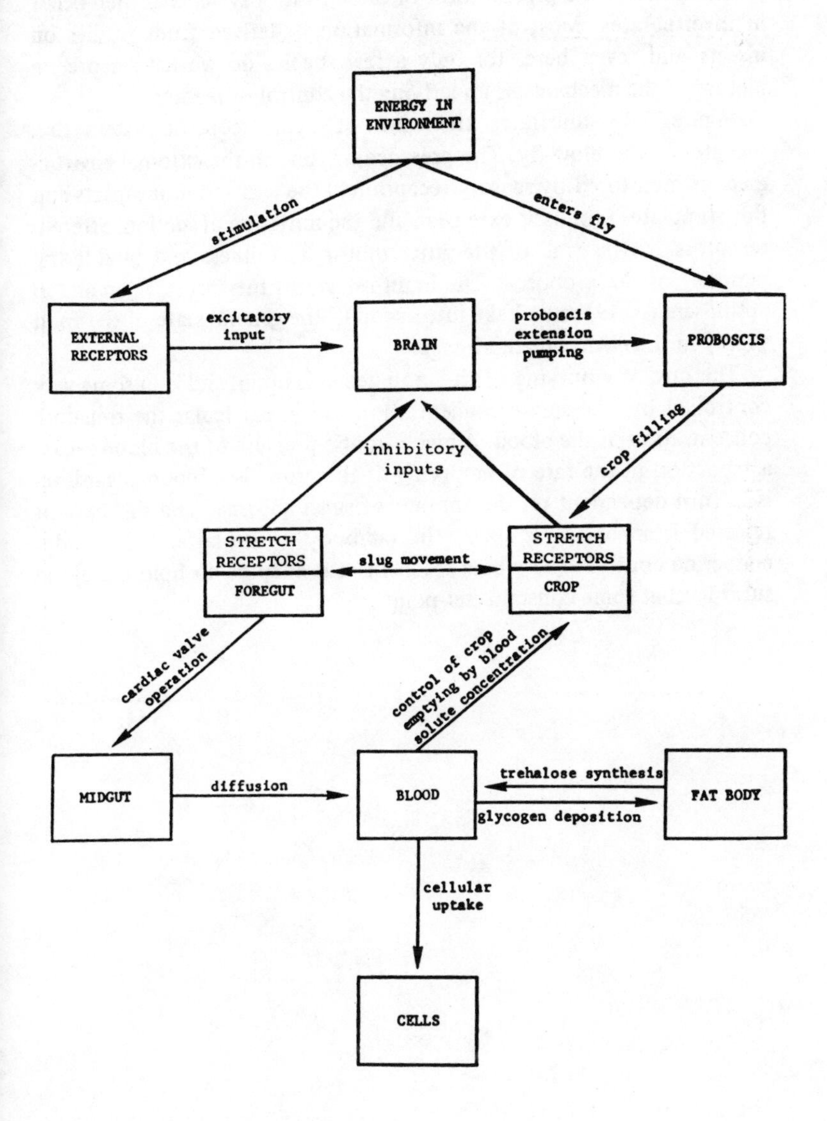

Source: With permission from Gelperin (1971).

loops in the neural and endocrine control systems of animals. All this can be deduced from first principles but there is, as yet, little experimental data on the precise form of these control systems as they occur in invertebrates. Most of the information is derived from studies on insects and, even here, for only a few species do we have a precise picture of the mechanisms underlying the control of feeding.

Figure 2.18 illustrates the kind of neural control system that operates in the blow-fly. The presence of food in the external environment is monitored by sensory reception in the legs and mouthparts and this stimulates proboscis extension and the initiation of suction. Stretch-receptors in the wall of the gut monitor its fullness and inhibit the pumping of the proboscis. The brain integrates this information and in optimisers (p. 49) must 'take into account' the general state of the food supply in the environment at large.

The rate of emptying of the crop depends upon, and is in some way controlled by, the solute concentration and in particular the trehalose concentration of the blood. A high osmotic pressure of the blood causes a reduction in the rate of emptying of the crop. The blood sugar level is in turn dependent on the amount of sugar absorbed and the amount released from or taken up by the fat body – the latter being under endocrine control. The whole mechanism is adapted to hold the blood sugar level at some constant 'set-point'.

3 RESPIRATION

3.1 Molecular Basis

Absorbed energy is partitioned between two major pathways: synthesis (processes involved with building and replacing the tissues) and respiratory metabolism (Figure 3.1). The latter transfers some of the chemical, potential energy in food to high-energy, phosphate bonds (~P), usually in adenosine triphosphate (ATP) molecules, and these, or at least the energy that they store, are then used to power mechanical work (e.g. muscular contraction), chemical work (e.g. active transport) and synthesis itself. Before proceeding, however, the reader should observe some caution on terminology. 'High energy phosphate bond' does *not* refer to the bond energy of the covalent linkage between the phosphorus atom and the rest of the molecule. The phosphate bond energy is not localised but is a reflection of the energy content of the whole triphosphate molecule before and after its conversion to a diphosphate. The phrase 'high energy phosphate bond' is, nevertheless, widespread and will be used in what follows as a convenient shorthand.

Figure 3.1: The utilisation of input energy by heterotrophs.

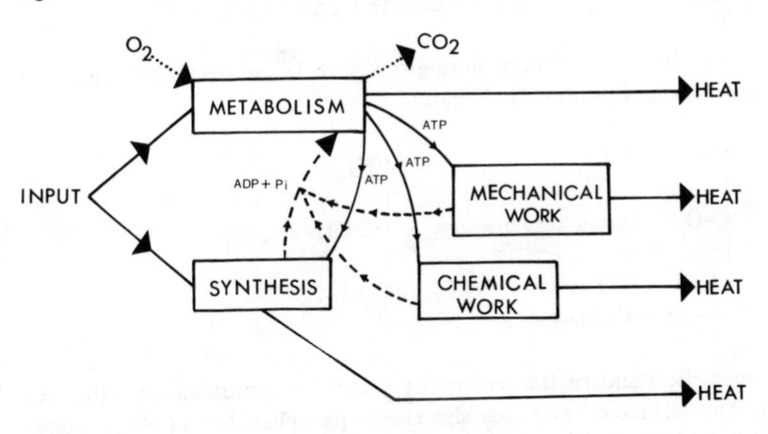

The maximisation principle would suggest that natural selection ought to have acted to reduce the costs of metabolism and hence to have maximised the efficiency of ATP-production and this we examine first before going on to consider how internal and external factors affect the rate and efficiency of the respiratory processes.

3.1.1 Glycolysis and the TCA Cycle

The metabolic pathways involved in the generation of ATP do not differ greatly from one phylum to another and are depicted in Figure 3.2. The main fuel is glucose, derived directly from the food or indirectly, by enzymatically-mediated transformations, from other molecules in the food or from glycogen and lipid stores. There are two main parts to this reaction sequence: glycolysis, which takes place in the cytoplasm of cells and may occur without oxygen, and the tricarboxylic acid cycle, which occurs in mitochondria and requires oxygen. Glycolysis generates ATP from ADP plus inorganic phosphate (P_i) by the direct involvement of substrates passing down the metabolic pathway, i.e. *substrate-level phosphorylation.* The TCA cycle generates reduced nicotinamide adenine dinucleotide (NAD_r) which is then used to convert ADP+P_i to ATP by donating electrons (and thereby becoming transformed to an oxidised state, NAD_o) to a transport system in which the final electron acceptor is oxygen. This is *electron transport phosphorylation.* Since, before the advent of green plants the earth's atmosphere was anoxic, glycolysis is thought to be more primitive than the TCA cycle.

Under anaerobic conditions we have the following balance sheet for a lactate-producing system:

$$\text{Glucose} + 2\text{ADP} + 2\text{P}_i \rightarrow 2\ \text{Lactate} + 2\text{ATP} + 2\text{H}_2\text{O}$$

Some NAD_r is produced during this process and is mopped up in the conversion of pyruvate to lactate:

$$\begin{array}{c} CH_3 \\ | \\ C{=}O \\ | \\ COO^- \end{array} \xrightarrow[\text{NADH+H}^+\ (\text{NAD}_r)\ \rightarrow\ \text{NAD}_o]{\text{Lactate dehydrogenase}} \begin{array}{c} CH_3 \\ | \\ HC{-}OH \\ | \\ COO^- \end{array}$$

Since the standard free energy of glucose is approximately -2881 kJ (-686 kcal) $mole^{-1}$ and each high energy phosphate bond (~P) is equivalent to 30.7 kJ (7.3 kcal) $mole^{-1}$ then the efficiency of conversion is:

$$\frac{2 \times 30.7}{2881} \times 100 = 2.13\%$$

Figure 3.2: Extremely simplified representation of anaerobic and aerobic respiration. Carbohydrates are converted to pyruvate and then lactate (or some other substance like octopine) in the anaerobic sequence. Pyruvate is successively degraded in the aerobic, tricarboxylic acid cycle.

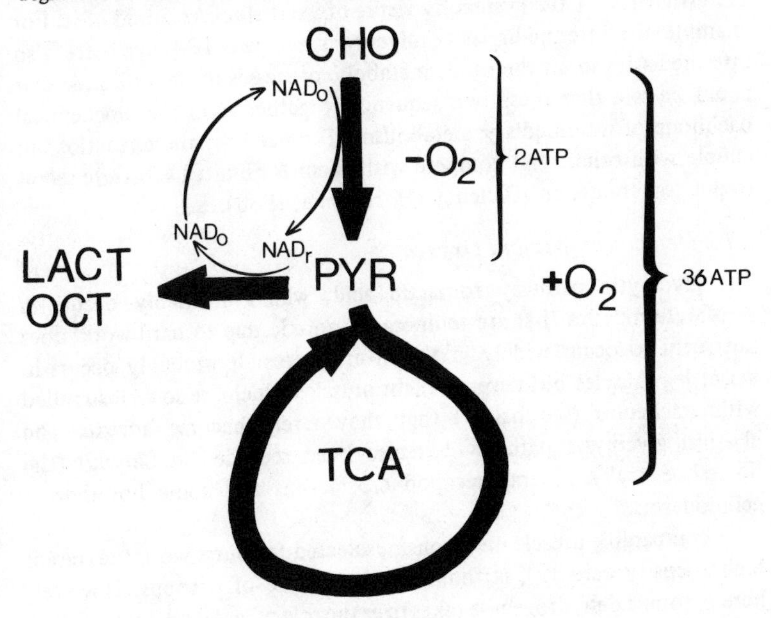

In other words the primitive anaerobic system is not very efficient; most energy going into the process is lost as heat.

In the presence of oxygen, the balance sheet becomes:

$$\text{Glucose} + 36\text{ADP} + 36\text{P}_i + 6\text{O}_2 \rightarrow 6\text{CO}_2 + 6\text{H}_2\text{O} + 36\text{ATP}$$

and the efficiency is therefore:

$$\frac{36 \times 30.7}{2881} \times 100 \simeq 38\%$$

Thus with the advent of green plants and oxygen it became possible to use the glucose more completely and more efficiently but not, perhaps, as efficiently as might have been expected. Again most of the input energy is lost as non-biologically useful heat. There are two possible reasons for this. First, the make-up and efficiency of the system we find now may have been fixed by chance happenings at the origin of life. It was presumably just these initial conditions which determined that carbon should form the chemical basis of life, that the fuel for metabolism should be carbohydrates and that phosphate molecules should be used as energy carriers. Secondly, constraints may be imposed on the efficiency of the system by virtue of what else is required of it. For example the intermediaries of glycolysis and the TCA cycle are also intermediaries in all the other metabolic processes of the cell. Indeed it could be said that these two sequences together form the biochemical backbone of intermediary metabolism. The need for these reactions to couple with others and compete with them for limited substrate seems to put constraints on efficiency (Hochachka, 1980).

3.1.2 Metabolism Without Oxygen

The glycolytic pathway to lactic acid, which invariably occurs in vertebrate muscles that are temporarily anoxic due to hard work, does not seem to occur widely in the invertebrates. It probably occurs in insect leg muscles but never in flight muscles which are so well supplied with tracheoles (see below) that they rarely become anoxic. The classical glycolytic pathway has also been recorded in *Limulus* (the horseshoe crab), several decapod crustaceans and some holothurian echinoderms.

A comparable muscle metabolism, selected for burst work (i.e. rapid, high-intensity activity), is found in the mantle of octopus. However, here octopine dehydrogenase takes over the role of lactate dehydrogenase and arginine phosphate takes over the role of creatine phosphate. In vertebrates the latter acts as a store accepting ~P from ATP in periods of relaxation and delivering it up during times of hard work:

$$\text{Phosphocreatine + ADP} \underset{\text{relaxation}}{\overset{\text{hard work}}{\rightleftharpoons}} \text{Creatine + ATP}$$

Hence, in this system creatine accumulates during burst work. Arginine phosphate operates in exactly the same way as phosphocreatine but arginine, unlike the creatine, is then involved in the reduction of pyruvate, and free arginine does not accumulate:

$$\text{Pyruvate + Arginine} \xrightarrow[\text{NAD}_r \curvearrowright \text{NAD}_o]{\text{octopine dehydrogenase}} \text{Octopine}$$

Organisms, such as benthic bivalves and some endoparasites which live for long periods in conditions of low or zero oxygen tension, use a different method of metabolism. The essence of this technique is that it retrieves energy from the NAD_r generated in the initial glycolytic pathway by using electron transport phosphorylation but with some molecule other than oxygen as the final electron acceptor. A simplified version of how this works is given in Figure 3.3. Phosphoenolpyruvate, the molecule before pyruvate in glycolysis, is converted to oxaloacetate and then fumarate, both intermediaries in the TCA cycle. The latter, rather than oxygen, acts as the final electron acceptor in the system of phosphorylation described above. It initially accepts electrons from NAD_r to produce NAD_o and to convert ADP + P_i to ATP. In a sense the TCA cycle is thereby put in reverse and some would argue that it was from this kind of beginning that the TCA cycle itself evolved.

Depending on the exact pathways employed between five and eight ATPs may be formed by this method, increasing the efficiency of the anaerobic system from 2 to between 6 and 8 per cent. This being the case, the question immediately arises as to why this method of metabolism has not replaced the lactate/octopine-producing one. One possible answer is that the more efficient succinate system produces ATP more slowly than the less efficient lactate system (Table 3.1) and this also applies to the octopine system etc. Hence the latter may be more advantageous when sustained output is required over short periods of anoxia as with over-worked muscle. Alternatively long-term anoxibiosis favours the more efficient system.

A good review of the molecular basis of metabolism is given in Lehninger (1973). Strategies associated with anaerobiosis are discussed in Hochachka and Somero (1973), in de Zwaan and Wijsman (1976) and more recently in Hochachka (1980).

3.2 Oxygen Availability and Uptake

Aerobic metabolism depends on oxygen being made available to respiring tissues and carbon dioxide being removed from them. Early this century many physiologists thought that some tissues could secrete oxygen and carbon dioxide even against a pressure gradient. All the evidence available to us now, however, suggests that respiratory gases always move across an interface by passive diffusion. Some animals rely on diffusion alone for the transport of oxygen and carbon dioxide throughout the tissues. Others use circulating blood for gas transport but gases still have to diffuse into and out of the blood at respiratory surfaces and at the interface with metabolising tissue. Yet others have air-filled tubes which connect the tissues with the external environment, but

Figure 3.3: Very simplified representation of anaerobic pathways. Most abbreviations are obvious and are elaborated further in the text. ETS = electron transfer system. etc = lactate and octopine and other end-products. Dotted line = TCA cycle.

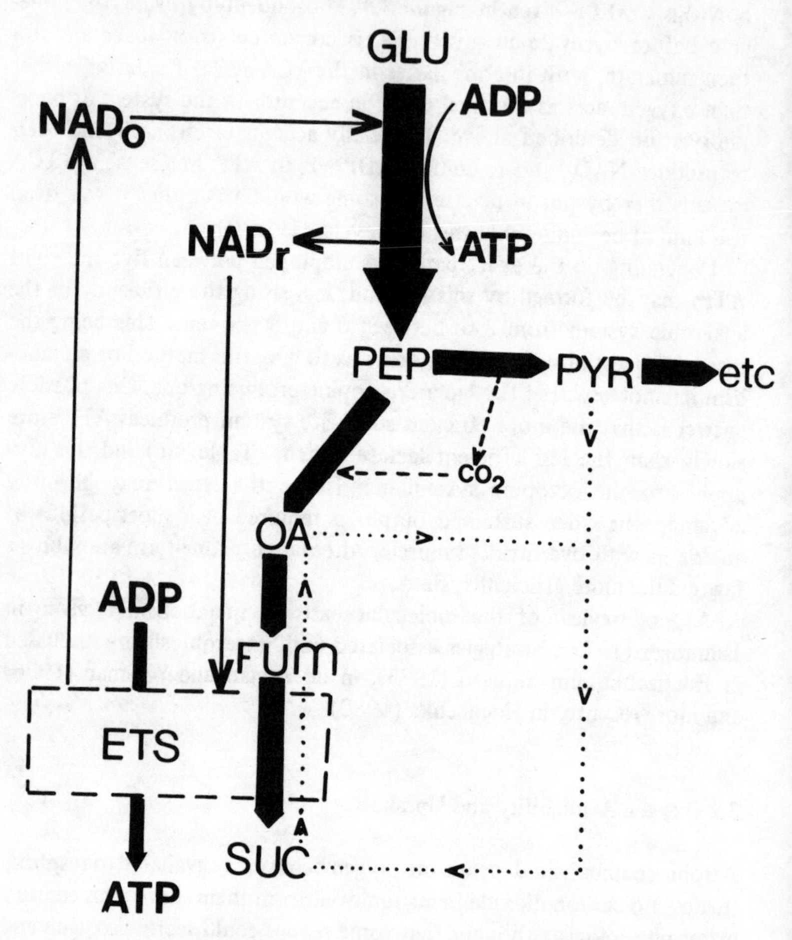

here again diffusion is the main method of movement of gas molecules in the tubes. Hence diffusion is a fundamental process of gas exchange and hence of aerobic respiration.

A gas travels by diffusion from regions where its partial pressure is high to regions where it is low. The partial pressure of a gas in a mixture of gases is the pressure it would exert if it alone occupied the whole mixture. Thus for oxygen:

Table 3.1: Relative efficencies and power output of standard anaerobic metabolism in trout muscle and the more elaborate pathways found in oyster muscle.

	Approx. Efficiency %	Approx. ATP Yield μmol $s^{-1} g^{-1}$ glucose
Trout muscle	2	13
Oyster heart muscle	5-8	1

Source: Hochachka (1976).

$$PO_2 = PF_g$$

where PO_2 = partial pressure of oxygen; P = total pressure of gas, and for air at normal barometric pressure = 760mm Hg (NB: the pascal is now the internationally accepted unit of pressure, but to date most biological research has used mm Hg – hence this is retained here – there are 133.32mm Hg per Pa); F_g = the fraction of oxygen in the gas, i.e. 20.95 per cent of air is oxygen. Hence in air at sea level, PO_2 = 760 (20.95/100) = 155mm Hg. As P reduces (e.g. with altitude) so does PO_2. For a liquid the fractional concentration of oxygen within it (F_l) is dependent upon the PO_2 of the gas phase applied to it, so $F_l = (\alpha PO_2)/P$ where α, a proportionality constant, depends on the gas, the solvent, the amount of other solutes (e.g. salt) and temperature. For example, at 15°C α for O_2 in sea water is 3.5, so 100 volumes of seawater will take up to (155/760) 3.5 or 0.7 volumes of oxygen at saturation. α is lower for salt-water than for freshwater and reduces with increasing temperature. Since, under standard conditions, α and P are constants, PO_2 is as good a measure of oxygen availability in liquid as it is for the gas phase.

Now the rate of diffusion of oxygen through the tissues depends not only on PO_2 gradients but also on the properties of the tissue; i.e. substances vary in the extent to which gases can pass through them and this is often expressed as a diffusion coefficient. Given realistic values for the diffusion coefficient of animal tissues and their oxygen demand (deduced from metabolic rate) it can be calculated that the distance between the tissue and a respiratory surface can be no more than 1 mm and this puts severe limitations on the size that can be attained by animals which depend on diffusion alone for the supply of oxygen. Nevertheless, some invertebrates with solid bodies do achieve considerable sizes. For example, some terrestrial flatworms may reach a size of more than 10 cm long and 1 cm wide and this they do, as the name implies, by flattening of the body so that the diffusion requirements are maintained. Similarly many jellyfish and anemones grow very big but rely exclusively on diffusion. The ectoderm and endoderm of these animals consist of a thin layer of cells in direct contact with water on the outside or circulating within the gastrovascular cavity. The thicker mesogloea, between these two layers, contains few cells, at least in jellyfish, and requires little oxygen. In the anemones, which have a more cellular mesogloea, this is less thick and usually within 1 mm of the external water or the water of the central cavity.

One way round the limitation imposed by diffusion is the evolution of an internal convection or circulation system. This increases the rate of transport of dissolved gases through the body and, by rapidly removing oxygen from the respiratory surface, maintains a steep PO_2 gradient and this increases the rate of intake. From the point of view of economisation in the making and maintenance of the blood system and for the sake of the efficiency of circulation, the vessels should be small and should be directed to specific places. But the smaller the vessels the greater the resistance and turbulence within them and the greater the energy required to maintain flow. Circulatory systems will represent some kind of balance between these opposing forces which may be determined as much by the organisational units which have been available in the evolution of the system (historical constraints) as by ecophysiological constraints. In fact two main kinds of system are found in the Invertebrata; the open system with a large haemocoel (expanded vessel containing blood) which is well developed in the Arthropoda and Mollusca and the closed system of arteries and veins found in the Invertebrata: the open system with a large haemocoel operate at low pressure and with a 'weak heart' whereas closed systems operate at higher pressure and require a more active and more muscular pumping system. On the other hand, the open systems contain a greater volume of blood and probably give less efficient circulation. There are intermediates. For example, the adaptation of isopod crustaceans to aquatic, amphibious and terrestial habitats is associated with increasing efficiency of the circulatory system. This culminates in land isopods, which have relatively larger and more muscular hearts than other crustaceans and strictly defined, often vessel-like lacunae which contribute to an almost complete circulatory system. The latter ensure a better regulated and more efficient transport of blood. Insects, of course, though having an open system, do not make much use of it in the transport of oxygen. Instead they have evolved a system of air-tubes,

Figure 3.4: Transverse section through crayfish (right-hand figure) and earthworm (left-hand figure) to illustrate open and closed blood systems (dotted areas) respectively. h = heart, D = digestive gland, G = gut, H = haemocoel, M = muscle, C = coelom, n = nerve cord.

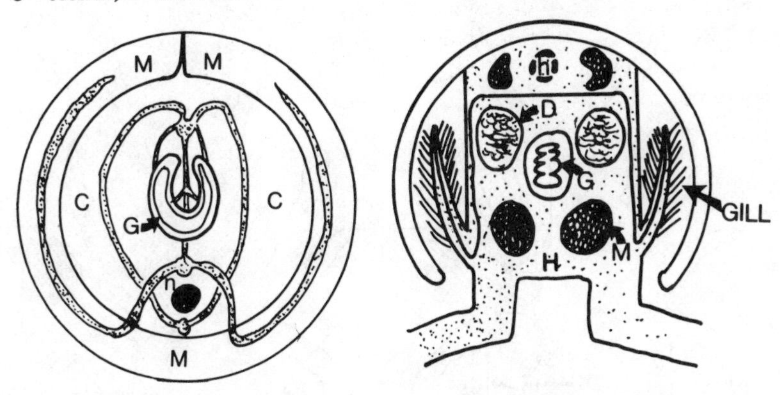

the tracheae, which allow oxygen to enter openings, the spiracles (or in aquatic forms from the gills), and pass directly by diffusion to the respiring tissues. An extensive and elaborate system of tracheae ensures that all the cells are provided for (Figure 3.5). Perhaps this system has evolved in response to the inefficiencies associated with the open, arthropodan blood system.

As well as depending on rate of flow, the effectiveness of a blood system in supplying metabolising tissues with oxygen depends upon the amount of oxygen it can carry per unit volume. Primitively, bloods are colourless and similar in composition to seawater. Such saline media can carry comparatively small amounts of gases and the bloods of many invertebrates contain oxygen-carriers – the so-called respiratory pigments – which are proteins capable of binding with oxygen. In order of commonness these are: haemoglobin (red when oxygenated); haemocyanin (blue when oxygenated); haemerythrin (pink when oxygenated); chlorocruorin (green when oxygenated). Haemoglobin is particularly characteristic of annelids and entomostracan crustaceans; haemocyanin occurs in many arthropods and molluscs; haemerythrin occurs in some polychaetes, sipunculids and priapulids; chlorocruorin occurs in several polychaetes.

The total amount of oxygen carried by the blood depends on the quantity of pigment it contains. However, an important property of a pigment for the animal is the extent to which it can take up and yield oxygen at different levels of PO_2. Figure 3.6 illustrates this property for several polychaete pigments as so-called oxygen dissociation curves.

Figure 3.5: Tracheal system of a terrestrial insect with spiracular openings (left) and an aquatic insect with tracheal gills (right). Spiracles and gills supply wide-bore tubes – the tracheae – and these lead to fine-bore tubes – the tracheoles – which supply the tissues.

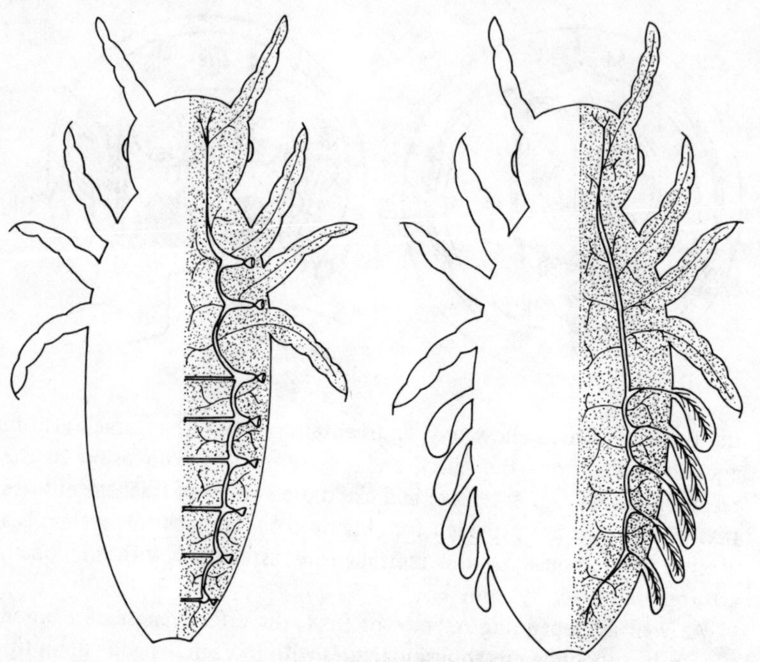

Source: Illustration by L.J. Calow, modified from Wigglesworth (1974).

These curves are more or less sigmoid in shape but differ in slope. The curve for *Arenicola* (measured at pH 7.5 and 19°C) indicates that its haemoglobin reaches a high degree of saturation at low PO_2, which means that it is capable of picking up oxygen at low levels of availability. *Nephthys* has one kind of haemoglobin in its blood and another in its coelomic fluid. Both (as measured at pH 7.4 and 15°C) are less good at picking up oxygen at low PO_2 than the haemoglobin of *Arenicola*. The chlorocruorin of *Sabella* (measured at pH 7.35 and 26°C) can pick up little oxygen at low PO_2 and functions best in well-oxygenated circumstances. These properties are often summarised as the PO_2 required to effect 50% saturation (i.e. P_{50}). Thus the P_{50} for *Arenicola* will be less than that for *Sabella* and the values for *Nephthys* will be intermediate. Some attempt has been made to give differences of this kind an adaptive explanation, but to date these have been sketchy. Thus the dissociation curve for *Arenicola* may aid it to obtain oxygen from the

Figure 3.6: Oxygen-dissociation curves of the haemoglobin of *Arenicola marina* at pH 7.5 and 19°C (A) and *Nephthys hombergii* at pH 7.4 and 15°C (B – vascular, C – coelomic fluids) and the chlorocruorin of *Sabella spallanzanii* at pH 7.35 and 26°C (D).

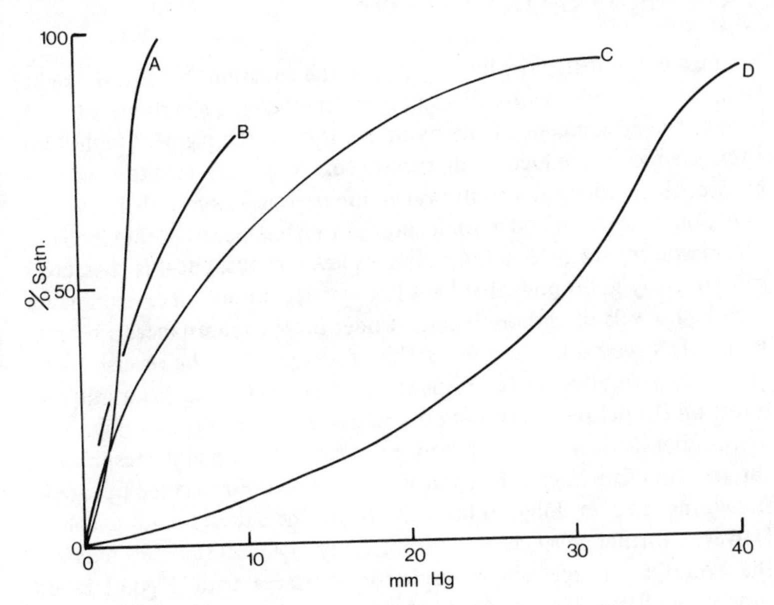

Source: Data from Jones (1972).

hypoxic muds in which it lives, particularly between tides. *Nephthys* also lives in a similar habitat but may be more capable of anaerobiosis between tides. *Sabella* lives sub-tidally and extends its crown of tentacles into currents of well-oxygenated seawater. For more detailed information see Jones (1972).

It must also be noted that the shapes of the dissociation curves, and hence the P_{50}s, are sensitive to environmental conditions (hence the need to specify the pH and temperature under which the annelid curves were determined above). For example the oxygen-affinity of pigments often reduces with increasing temperature and reducing salt concentration of the blood. Hydrogen ion concentration or pH also has a pronounced influence – known as the Bohr effect – which is particularly important physiologically. Most pigments show a normal Bohr effect in which a reduction in pH causes a decrease in P_{50} (curve shifts to the right) but some show the reverse of this. The physiological significance of the normal Bohr effect relates to the way pigments interact with hydrogen ions produced by respiring tissues. These influence the dissociation of CO_2 at the tissues according to the simple equation:

$$CO_2 + H_2O \rightleftharpoons H_2CO_3 \rightleftharpoons H^+ + HCO_3^-$$

If a pigment mops up the hydrogen ions, the equation shifts to the right and more CO_2 can be carried away from the tissue as bicarbonate in the blood. The attachment of the hydrogen ions to the pigment molecules alters their shape, reduces their affinity for oxygen and results in oxygen release. Hence there is a reciprocal action between oxygen delivery, carbon dioxide removal and gas transport during high levels of metabolism. The reason for the reverse Bohr effect is less obvious, but it is characteristic of many gastropods that have low activity but live in environments with high levels of carbon dioxide. Under these circumstances a normal Bohr effect would hinder the uptake of oxygen, but the reverse effect facilitates saturation at the respiratory surface while having a minimal effect on the delivery of oxygen to the tissues.

Another limitation on respiratory uptake is the size of the respiratory surface. An unspecialised body surface is adequate for oxygen uptake in flatworms and in long, thin nemerteans, nematodes and annelids. However, further advances in size or activity in aquatic animals required the evolution of specialised respiratory surfaces with a good blood supply. Similarly, the invasion of land, which brought with it the need for water conservation and hence an impervious cuticle, also depended on the evolution of restricted and therefore specialised respiratory surfaces. A stylised illustration of the main kinds of respiratory organ is given in Figure 3.7. In general, respiratory organs evaginate in aquatic organisms but invaginate in terrestrial ones. Clearly, respiratory organs must not consume more oxygen in maintenance metabolism than they generate and this will tend to limit the size of the organ as will the dangers of water loss in terrestrial organisms.

One other mechanism which assists in oxygen uptake is the ventilation of the respiratory surfaces. This occurs in animals with and without specialised respiratory organs. Arrangements range from the lateral cilia on the gill lamellae of bivalves to the muscular pumps servicing the anal respiratory trees of sea cucumbers. Tubiculous polychaetes depend on a ciliary current or peristaltic movements. Crustacea commonly depend on undulating appendages. Irrigation often increases with reducing PO_2 and may contribute to a regulated oxygen uptake despite diminishing supplies (see Section 3.3). Once again active ventilation is associated with respiratory costs (as is all physical activity; see Section 3.6), the

Figure 3.7: The main kinds of respiratory system found in invertebrates.

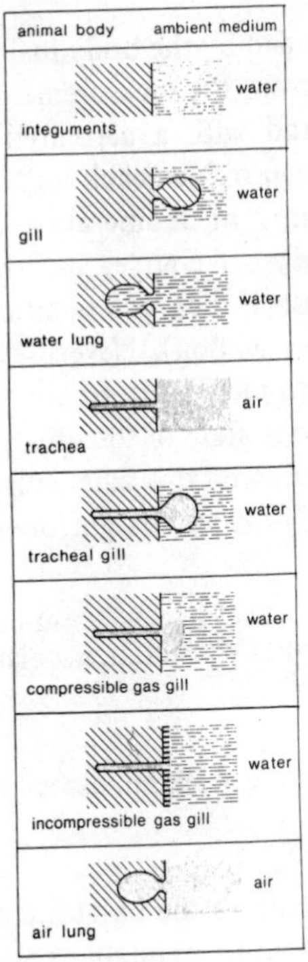

animal body / ambient medium	
integuments (water)	As in Platyhelminthes
gill (water)	Found in some Crustacea and Mollusca; also book-gill of *Limulus* and podia of some Echinodermata
water lung (water)	Holothurian respiratory tree and rectal respiration of some insect larvae
trachea (air)	Tracheae and book-lungs and tracheal lungs of Arachnida
tracheal gill (water)	Aquatic Insecta
compressible gas gill (water)	Physical gill of some aquatic Insecta
incompressible gas gill (water)	Plastron of some aquatic Insecta
air lung (air)	Pulmonate molluscs

Source: With permission from Dejours (1981).

precise magnitude of which has not been determined for invertebrates. Clearly, however, there is likely to be a diminishing return from such activity and as soon as the ventilation uses more oxygen than it creates it becomes uneconomic and should cease. Precise limits are likely to vary from situation to situation and have yet to be determined for any invertebrate. For fishes, some consideration is given to this problem by Hughes and Shelton (1962) and Jones (1971).

A last question can be asked: at what PO_2 does aerobic metabolism switch to the anaerobic pathways? Actually, we now find that there is no low, critical P_{O_2} at which switching occurs and even at high P_{O_2} some anaerobiosis may be taking place. This is an active area of research and no one yet knows the precise extent to which anaerobic mechanisms contribute to total metabolism at different PO_2s in invertebrates.

3.3 Levels of Metabolism

The easiest way of measuring the energy expended by the body in doing physical and chemical work (Figure 3.1) is from the measurement of oxygen uptake. Under aerobic conditions and with a carbohydrate substrate, approximately 21 J of heat are given off for each millilitre of oxygen inspired under standard conditions of temperature and pressure. Ease does not always equal accuracy, of course, and these measurements will underestimate metabolic costs if there is a significant amount of anaerobic respiration (see previous section). Nevertheless, since oxygen uptake is what has been determined most frequently, there has been a tendency to define the metabolic state of the organism in these terms. Invertebrate biologists have borrowed this scheme from mammalian physiologists (e.g. Brody, 1945). The most straightforward classification is as follows:

Standard metab. + Metab.associated with growth + Metab.associated with spontaneous activity = Routine metab.

Routine metab. + Metab. associated with feeding + Metab. associated with overt activity = Active metab.

In principle standard metabolism is equivalent to the basal metabolism of homeotherms and is that energy required to maintain existing tissue. Rarely is it possible that there is no synthetic activity or physical work going on, however, and so the lowest rate of oxygen uptake that is recorded in respiratory studies is often referred to as resting or quiescent metabolism.

The difference between the oxygen uptake in the resting and active condition is sometimes referred to as the *scope for activity* (Fry, 1958). This is a measure, in terms of oxygen, of the energetic potential of the system for 'biological work'. Each level of metabolism may respond differently to changes in external conditions and this means that the scope for activity varies with external conditions. For example, the capacity for the mussel, *Mytilus edulis*, to regulate its rate of oxygen consumption during reductions in environmental PO_2 (environmental hypoxia) varies with level of metabolism (Bayne *et al.*, 1976). When mussels are starved their oxygen consumption rates approximate to a standard level and are linearly dependent on PO_2. On the other hand routine metabolism at first resists reductions in PO_2. Standard metabolism is therefore said to conform to PO_2 and routine metabolism is said to show a regulatory response; both are illustrated in Figure 3.8. (Some species regulate and others conform at all levels of metabolism; a list is given in Table 3.2.)

Figure 3.8: The response of routine (R) and standard (S) metabolism (Vo_2 = ml O_2 hr^{-1}) to Po_2 in mm Hg. Scope is the difference between R and S.

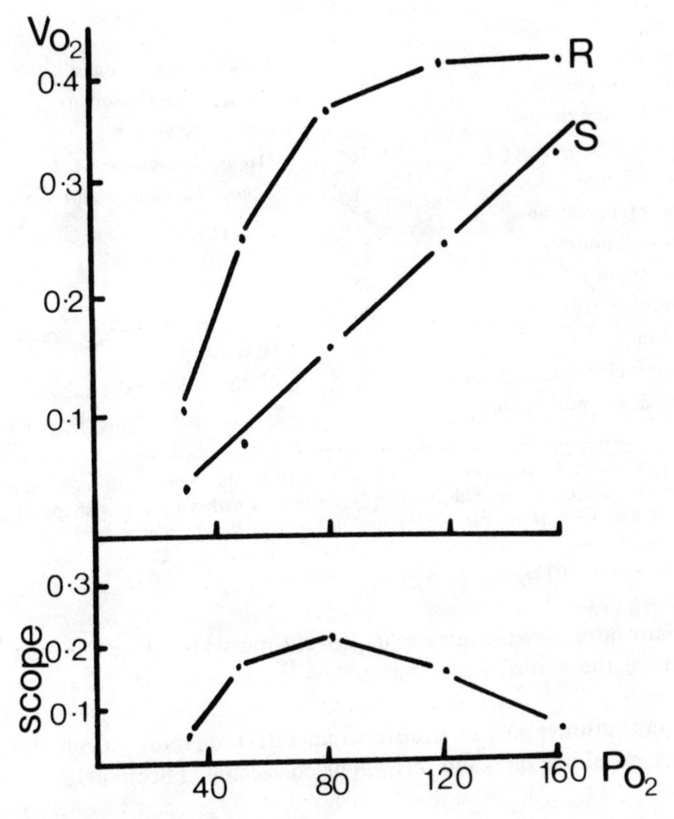

Source: After Bayne *et al.* (1976).

The numerical differences between the curves in Figure 3.8 are a measure of the scope for activity of the mussel. It increases in a slightly hypoxic environment to a maximum at 80 mm Hg PO_2 and then declines to the point where routine and standard curves coincide. Fry (1958) identified an incipient limiting tension (P_C) as the point at which the scope for activity begins to decline (for the mussel c. 80 mm Hg) and an incipient lethal tension as the point at which both routine and

Table 3.2: Examples of regulators and conformers. Tc is the critical Po_2 (mm Hg) at which regulation breaks down and regulators become conformers.

Regulators	Tc*	Conformers
Paramecium (ciliate)	50	*Spirostomum*
Tetrahymena (ciliate)	2.5	
Trypanosoma (flagellate)		
Pelmatohydra (hydrozoan)	60	Various sea anemones
Aurelia (scyphozoan)	120	*Cassiopea* (scyphozoan)
Tubifex (oligochaete)	25	*Nereis* (polychaete)
Lumbricus (oligochaete)	76	*Erpobdella* (leech)
Urechis (gephyrean)	70	*Sipunculus* (sipunculid)
Mytilus (lamellibranch)	75	
Helix (pulmonate)	75	*Limax* (pulmonate)
Loligo (squid)	45	
Cambarus (crayfish)	40	*Limulus* (king crab)
Uca (crab)	4	*Homarus* (lobster)
Cloeon (ephemerid nymph)	30	*Baetis* (ephemerid nymph)
Hyalophora (moth larva)	25	*Tanytarsus* (chironomid larval)

* The values of Tc must be accepted with caution since the transition from regulation to conformity is seldom sharply defined. Furthermore the shape of the oxygen uptake/tension curve is often influenced by degree of acclimation to abnormal Po_2 s.

Source: Jones (1972).

standard rates have the same value (for the mussel c. 20 mm Hg). Below this level the animal is doomed unless it can make use of anaerobic metabolism.

Temperature also has a differential effect on both standard and routine metabolic rates and this will be considered in Section 3.7.

3.4 Routine Metabolism and the Effect of Body Size

As yet we know little about the relative involvement of the costs of chemical work, physical work and the costs of synthesis in routine metabolism. Indeed it is more than likely that different measuring techniques themselves differentially affect the several aspects of the routine metabolic rate. Despite much variability, however, we do know that bigger animals respire more in absolute terms but less relative to their individual mass than smaller animals (e.g. Figure 3.9). That is to say, routine respiratory rates, as measured by oxygen uptake, increase but at a reducing rate with body size and this is represented by the following equation:

$$\text{Resp. rate} = a(\text{Weight})^{b}$$

where: a and b are constants and b is usually less than 1. Dividing throughout by weight gives:

$$\text{Resp. rate per unit weight} = a(\text{Weight})^{b-1}$$

Taking logarithms gives:

$$\log(\text{Resp. rate}) = k \text{ (or } \log a) + b \log(\text{Weight})$$
$$\log(\text{Resp. rate per unit weight}) = k + (b-1) \log(\text{Weight})$$

The constant, b-1, is sometimes referred to simply as b but to avoid confusion Davies (1966) has suggested that it should be denoted as b'.

The above equations not only describe the relationship between oxygen uptake and weight on an intraspecific basis, but they also describe it on an interspecific basis. A number of authors have demonstrated a linear relationship between log (Resp. rate) and log (Live weight) for a wide variety of unicellular and multicellular organisms ranging in size from protozoans to elephants and encompassing invertebrates. Regression coefficients, i.e. values of b, from two major studies are summarised in Table 3.3. Some authors have tried to explain these logarithmic equations on the basis of the geometric relationships between respiratory surfaces (which are assumed proportional in size to body surface area) and the mass of respiring biomass (which is assumed proportional to body live-weight). But this predicts a value for b of 0.67 and therefore for b' of 0.33. If it is assumed, for example, that an animal can be considered simply as a collection of cells, then it is clear that the weight of the cells is directly related to their volume and that the latter is proportional to the cubic power of their diameters. Alternatively the surface area of the cells is directly proportional to the square of their diameters. Hence cell surface is proportional to the cube root of weight raised to the power of 2, or $M^{2/3}$ (= $M^{0.67}$). Moreover, it is assumed that divergence from this simple geometric prediction in metazoa is due to the more complex respiratory surfaces found there and the supplementation of diffusion by ventilation and circulatory systems (see above). In other

Figure 3.9: As they get bigger animals respire more – because they contain more metabolising tissue. However, as is shown in these graphs the oxygen consumption per unit mass of tissue *reduces* with increasing size (expressed here in terms of nitrogen content) of animals. These relationships conform to fairly predictable and easy-to-define mathematical functions.

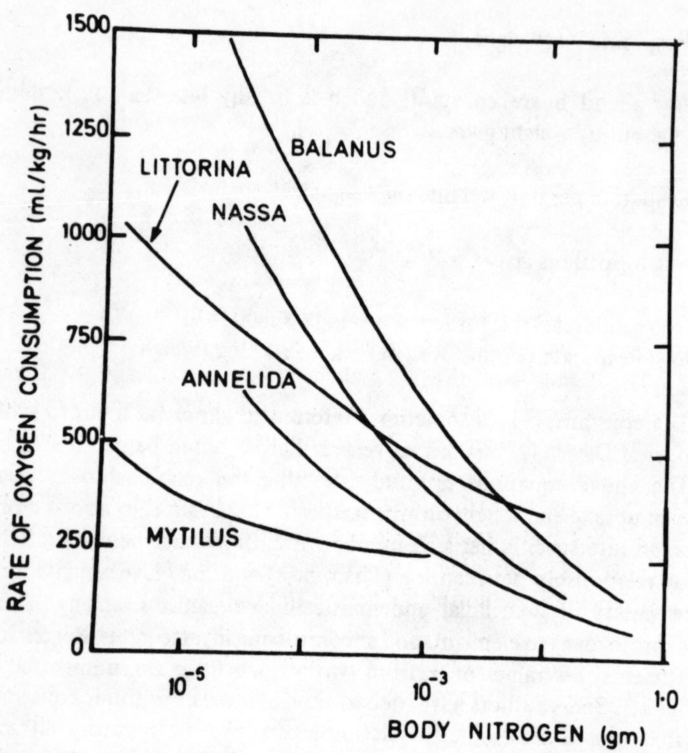

Source: After Newell (1970).

words the geometrical and metabolic properties of whole organisms are not just equal to the sum of the properties of their constituent cells.

However, there are a number of pieces of evidence which argue against these simple geometric arguments:

(1) There is much variation in the value of b even between closely related species of similar geometrical shapes; for example, b in woodland gastropods ranges from near zero to near one dependent on species (Mason, 1971). There is also some variation within-species dependent on external conditions; for example, temperature, oxygen concentration and food supply may all

Table 3.3: The values of the slopes of the regression lines relating the logarithms of respiratory rates and body mass (live weight) as calculated by two authors; i.e. b-values.

	Hemmingsen (1960)	Zeuthen (1970)
Unicellular ectotherms[a]	0.756 ± 0.021	0.70
Multicellular ectotherms[a]	0.738 ± 0.0095	0.80-0.95
Multicellular endotherms[b]	0.739 ± 0.010	0.75

Notes: a and b defined below, p. 87. See also Phillipson (1981).

influence the value of b in many invertebrates (Newell, 1970). Such variability would not be expected if simple geometric constraints were operating, since these depend only on shape and this is not influenced by external conditions.

(2) The double logarithmic size dependency of oxygen uptake persists even in tissue homogenates from the winkle, *Littorina littorea,* so the relationship does not depend on intact membranes and respiratory surfaces (Newell and Pye, 1971).

(3) The b-value for triploid *Drosophila,* with large cells and therefore a small cell-surface-to-volume relationship, is the same as for normal diploid individuals with smaller cells and a larger surface-to-volume relationship (Ellenby, 1953).

The basis of the relationships illustrated in Figure 3.9 is therefore complex and is unlikely to be attributable completely to geometrical constraints. Other factors must be involved, but exactly what these are remains to be elucidated.

3.5 Metabolism Associated with Feeding

Many animals experience a post-prandial increase in oxygen consumption which subsides shortly after the meal. This has often been referred to as specific dynamic action (SDA), specific dynamic effect (SDE) or simply the calorigenic effect.

Table 3.4 documents an SDE for the crustacean, *Macrobrachium rosenbergii* and gives a hint that different foods have different magnitudes of effect. This is also true for mammals where high protein diets have the most marked effects and high carbohydrate and fat diets have little effect at all. Arguing from this, many vertebrate workers have assumed that the elevated metabolic rate immediately after a meal derives from the expensive processes associated with catabolising excess proteins. The data for *Macrobrachium* do not support this conclusion, however, in that nitrogenous (NH_3) excretion does not differ significantly between diets. A possible alternative explanation is that SDE represents the cost of a surge in synthesis following the influx of fresh resources from the meal (Mitchell, 1962). The direct injection of amino acids into vertebrates provokes an SDE and suggests that the effect emanates from post-absorptive processes and not from the costs of eating and digesting the meal. The same may not be true for invertebrates since the costs of eating are sometimes considerable (Chapter 2). Bayne and Scullard (1977), for example, measured a post-prandial increase in the oxygen consumption of *Mytilus* which cost approximately 25 per cent of the food energy ingested. However, more than 80 per cent of this was attributable to the cost of filtering the food and less than 20 per cent to the cost of ammonia excretion.

Quantity as well as quality influences SDE in vertebrates; the greater the amount eaten, the greater the elevation of metabolism. Table 3.4 indicates that there was no significant correlation between these parameters in *Macrobrachium,* suggesting that SDE, here, was an all-or-nothing, threshold phenomenon. No SDE was recorded at all for *Calanus finmarchicus* (Marshall *et al.*, 1934) and *Daphnia* (Richman, 1958) but post-prandial elevations in oxygen consumption have been recorded for *Arcatia* species of copepod (Conover, 1956).

The reduction in oxygen uptake often associated with starvation in animals will include reduction in not only SDE, but also sometimes in active metabolism and possibly standard metabolism as well. These are part of the economy measures taken by animals in the face of food shortage and are discussed again in Chapter 7.

3.6 Active Metabolism

Movement involves muscular contractions and the beat of cilia and flagellae so that it is bound to enhance metabolism over the standard level. Table 3.5 shows that this is indeed the case for a variety of invertebrates with the active metabolic rate increasing between 1.4 times of the standard levels in the limpet, *Patella,* to approximately 100 times of the standard level in flying locusts.

It is *a priori* reasonable to expect a positive correlation between the

Table 3.4: SDE and excretion data for *Macrobrachium rosenbergii.*

Diet	Oxygen uptake before feeding ml. g^{-1} . hr^{-1}	Oxygen uptake after feeding ml. g^{-1} . hr^{-1}	% increase	Ammonia excretion after feeding mg. g^{-1} . hr^{-1}	Correlations for relationship of metab. rate and amount ingested	
					r*	p**
Tubificids (worms)	0.959	1.355	39.4	0.037	+0.247	>0.05
Cladophora (algae)	0.872	0.943	7.1	0.039	+0.003	>0.05
Purina (commercial food)	1.053	1.252	19.9	0.033	-0.130	>0.05

* Correlation coefficient
** Significance level – i.e. they are not

Source: Nelson *et al.* (1977).

Table 3.5: Comparison of active and standard rates of oxygen uptake in a variety of invertebrates.

Species	Temp. °C	Dry weight of animal mg	Standard rate μl/mg dry wt/hr	Active rate μl/mg dry wt/hr	Increase
Gammarus oceanicus	15	150-350	0.61	1.5	x 2.5
Arenicola marina	17.5	500	0.2	0.8	x 4.0
Patella vulgata	15	500	0.42	0.6	x 1.4
Actinia equina	15.5	100	0.5	18.0	x 36.0
Nephthys hombergii	15	100	0.12	0.8	x 6.6
Littorina littorea	15	100	0.25	3.0	x 12.0
Cardium edule	15	100	0.3	1.9	x 6.3
Balanus balanoides	14.0	100	0.25	1.2	x 4.8
Insects in flight	ambient	various	–	–	x100*

Source: Mainly from Newell (1970); * from Krogh and Weis-Fogh (1951).

speed of movement and the intensity of active metabolism, but it is not easy to envisage what precise form this relationship should take. Figure 3.10a shows a curvilinear relationship between oxygen consumption and flight speed in the desert locust, *Schistocerca gregaria*, but Figures 3.10b and c show linear relationships between oxygen consumption and activity for swimming in *Gammarus* (an amphipod) and crawling in a slug.

3.7 Effect of Temperature

Temperature can have two main effects on biochemical reactions and hence on metabolism:

(1) *Rate effects* – increasing temperature causes an increase in the average kinetic energy of reacting atoms and molecules and thereby causes an increase in rates of reaction. However, temperatures on earth are relatively low from this point of view and, without the influence of enzymes, fluctuations in earthly temperature would have little influence on biochemical reactions. Enzymes work by straining the bonds in reacting molecules such that they are made more susceptible to the influence of earthly temperatures.

(2) *Denaturation* – weak structural bonds abound in biochemical molecules. They influence the shape and function of macromolecules, particularly enzymes. Weak bonds are easily disrupted at physiological temperatures causing the macromolecules to become deformed and to lose their biological function. This is referred to as denaturation.

The relative importance of these rate and denaturation effects is illustrated in Figure 3.11 for a hypothetical enzyme-mediated reaction.

The influence of temperature on metabolic rate can be summarised by many mathematical functions. However, too much should not be read into these regarding the nature of the underlying processes; rather they are best treated simply as indices of the effects of temperature change. The simplest, most straightforward and most widely used index is the Q_{10} value. This is calculated as follows:

$$Q_{10} = \frac{R_1}{R_2}^{10/(t_1 - t_2)}$$

$$\text{or } \log_{10} Q_{10} = \frac{(\log_{10} R_1 - \log_{10} R_2)\ 10}{t_1 - t_2}$$

Figure 3.10: Effect of activity on respiratory rate: (A) *flight* of locust (here, resp. rate = kcal. kg^{-1} . hr^{-1} , speed = $m.s^{-1}$); (B) *swimming of Gammarus* (here, resp. rate = μl O_2 . 0.25 g^{-1} . hr^{-1} , speed = activity peak hr^{-1}) after 1 hr observation (circles) 4 hr observation (triangles); (C) *crawling* in the slug *Agriolimax columbianus* (here, resp. rate = $W.kg^{-1}$, speed = $mm.s^{-1}$).

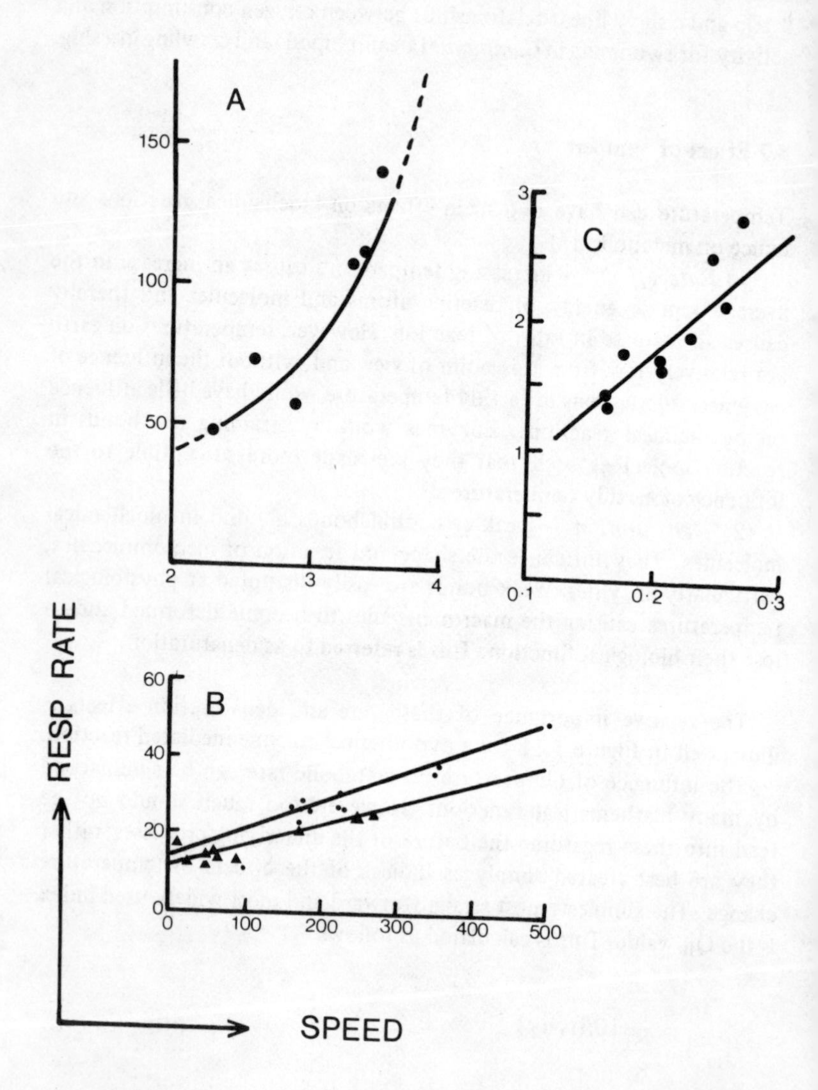

Source: (A) after Weis-Fogh (1954); (B) after Halcrow and Boyd (1967); (C) after Denny (1980).

Figure 3.11: Relationship between the rate of an enzymatically-mediated reaction (R) and temperature (T).

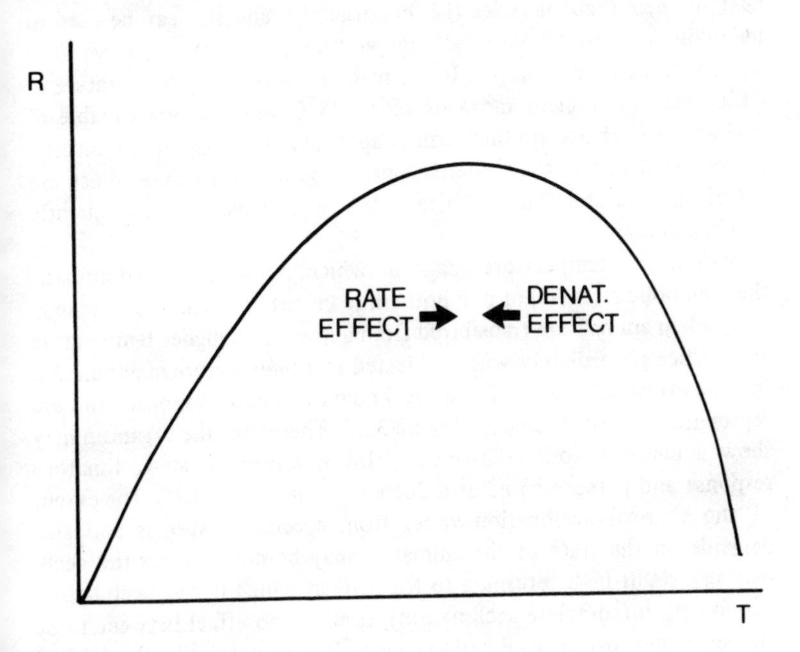

where R_1 and R_2 are metabolic rates at temperatures (°C) t_1 and t_2 respectively. Numerous Q_{10} values have been tabulated. In general, however, chemical processes have a Q_{10} which tends to two; i.e. there is a two-fold change in reaction rate for every 10°C change in temperature. Relative to chemical reactions, a Q_{10} of less than two denotes less sensitivity to temperature, a Q_{10} of one denotes insensitivity and a Q_{10} of more than two denotes greater sensitivity to temperature. The Q_{10} values of the metabolic rates of invertebrates stretch throughout this range.

Clearly, changes in ambient temperatures only influence metabolic rates if they cause changes in body temperature. This is the case in invertebrates but, within limits, is not true of mammals and birds. Invertebrates are therefore described as *poikilothermic* ('poikilo' = 'varied' in Greek) and mammals and birds are said to be *homeothermic* or *homoiothermic* ('homoio' = 'keeping the same' in Greek). A few invertebrates, however, such as those of the deep sea and antarctic waters, live at constant temperature and for this reason invertebrates are often described generally as *ectothermic* – denoting the source of body heat. Similarly, mammals are often described as *endothermic*, because they generate heat from within. Note again, however, that there are exceptions. For example, many insects generate large amounts of heat in their flight muscles (i.e. endothermy) and this can be used to maintain a constant thoracic temperature (i.e. homeothermy). The Sphinx Moth is an example. It can maintain a thoracic temperature of 41°C over an ambient range of 15 to 35°C and bees are capable of similar feats. Hence no one term is applicable throughout the invertebrates to describe their thermal properties. Nevertheless, most *are* poikilothermic and this is the term that will be used most frequently in what follows.

Within the temperature range in which rate-effects predominate, the metabolic rates (R) of poikilothermic invertebrates increase immediately when animals are transferred from a lower to a higher temperature and reduce immediately when subjected to a temperature manipulation in the reverse direction. These are known as acute responses and are represented as broken lines in Figure 3.12. Thereafter the organism may show a compensatory adjustment. This is known as an acclimatory response and is represented as a dotted line in Figure 3.12. The extent of the thermal acclimation varies from species to species and also depends on the state of the animal. It may be non-existent (no acclimation), result in R returning to the level at which it had been before manipulation (complete acclimation), result in an effect between these two extremes (partial acclimation), result in R overshooting the original value (over-compensation) or result in R diverging from its original value (reverse or negative acclimation).

Figure 3.12 illustrates two extreme cases of thermal acclimation. In (a), when animals living at low temperatures (or kept there in the laboratory) were transferred to warmer conditions they had a higher metabolic rate than animals which had been living (kept) at high temperatures. The rate of the low-temperature group gradually decayed to the lower rate giving partial, positive acclimation. The reverse occurred when animals were transferred to low temperatures. Hence the R/T curves linking the acute rates (broken lines) are approximately parallel and the metabolic rates of warm and cold acclimated animals measured at temperatures between the extremes would be expected to fall on these two lines. These lateral shifts in R/T curves are referred to as *translation.* Note that the Q_{10} values between acute rates will be greater than the Q_{10} values between acclimated rates over the same range. Such an acclimation response is thought to be adaptive because it allows conservation of energy at high temperatures and the generation of ATP at low temperatures so that body maintenance and vital activities

Figure 3.12: R/T curves (within the rate-effect component – Figure 3.11) to illustrate positive (a) and negative or reverse (b) acclimation. Spots represent rates. Broken lines indicate experimental manipulations. Dotted lines indicate direction of acclimatory adjustment. In each case the solid line links acclimated spots; note that this line is not as steep as the broken lines in (a) but is more steep in (b).

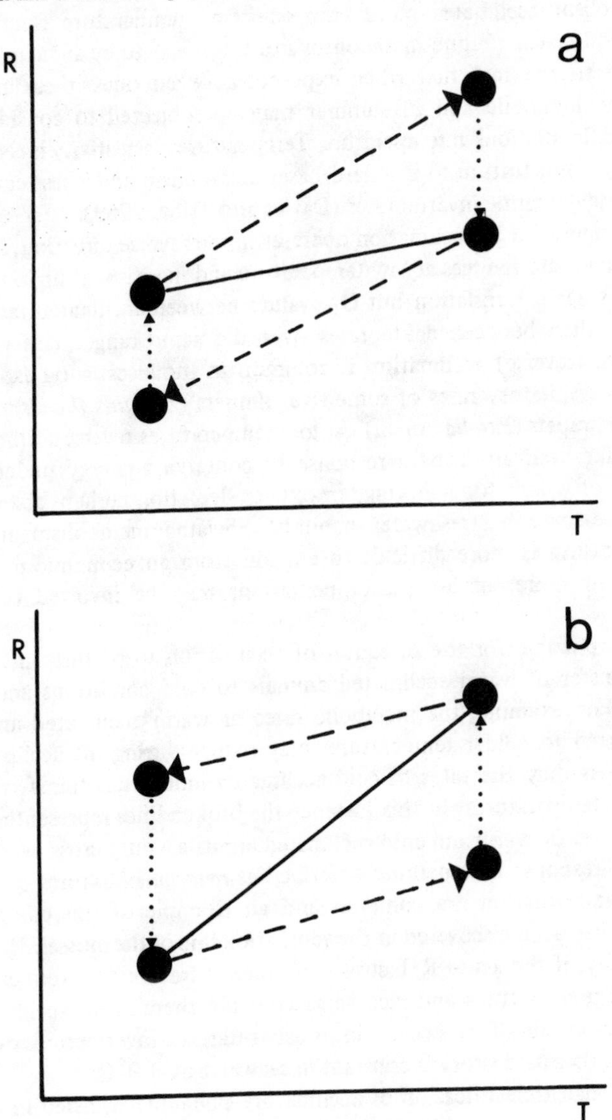

can be kept going. Partial positive acclimation is common in invertebrates (see Newell, 1970 and series of reviews in *Netherland Journal of Sea Research,* 7) and has been discovered in Platyhelminthes, Mollusca, Annelida and Echinodermata. In intertidal invertebrates the standard metabolism acclimates more completely to temperature fluctuation ($Q_{10} \triangleq 1$) than routine metabolism and this seems to be an adaptive response to the fact that when exposed, between tides, these animals become immobile and in summer may be subjected to considerable diurnal fluctuations in temperature. Temperature insensitivity is certainly not a general feature of standard/basal metabolism and is less common in subtidal, marine invertebrates (Davies and Tribe, 1969).

In Figure 3.13b acclimation operates in the reverse direction; i.e. the metabolic rate reduces at low temperature and increases at high. Hence, there is again translation but Q_{10} values between acclimated rates are greater than between acute rates over the same range. This partial, negative (reverse) acclimation is common in molluscs; good examples are the respiratory rates of some river limpets (*Ancylus fluviatilis*) and marine limpets (*Patella aspera*). At low temperatures reverse acclimation may, like aestivation, be a response to conserving energy under conditions of winter food shortage or oxygen-depletion perhaps associated with ice-cover in freshwater habitats. Elevated metabolism at high temperatures is more difficult to explain from an economic point of view and more subtle rate-compensations may be involved (Calow, 1975).

It is possible for the direction of acclimation to be the same after the transfer of warm acclimated animals to cold conditions and vice versa. For example, the metabolic rates of warm acclimated animals transferred to colder temperatures may reduce during the acclimation periods as may the rates of cold acclimated individuals transferred to warmer temperatures. In this instance the broken lines representing the acute rates of warm and cold acclimated animals would cross over. This type of response is sometimes described as *rotation.* Mixtures of translation and rotation are common and an example of this composite pattern has been discovered in the ventilation rate of the mussel, *Mytilus.*

Finally, if the acute R/T curves of animals transferred from cold to warmer temperatures and vice versa coincide there is no acclimatory response at all. This occurs in most antarctic invertebrates whose previous thermal history is constant in seawater at -1.9°C.

Systems of classification of acclimatory phenomena, listed in order of complexity and sophistication, have been devised by Precht (1958), Prosser (1973) and Alderice (1972).

At the biochemical level there are three major ways of adjusting enzyme activity in response to temperature-disturbance (Hochachka and Somero, 1973):

(1) *Quantitative strategy* – changes are effected in the concentration of pre-existing enzymes – e.g., in positive acclimation either more is produced or less is lost per unit time at low temperatures.
(2) *Qualitative strategy* – changes in types of enzymes synthesised.
(3) *Modulation* – by interacting with weak bonds the temperature change itself may alter the activity of the enzyme in the required direction.

The quantitative strategy depends in part on protein synthesis which is expensive and may therefore be self-defeating at low temperature; but adjustment may also be effected here by a reduction in protein turnover. To date there is no evidence to suggest that the qualitative strategy is more widespread. Modulation is almost certainly employed where rapid compensation is required.

It must also be noted that most of the research that has been carried out on the relationship between temperature and metabolic rate has been done at constant temperature. Yet in nature temperature is much more variable and this is particularly true in terrestrial, shallow-aquatic and inter-tidal habitats. From the little information available it would seem that the response at variable temperatures certainly differs from that at constant temperature but it is too early to generalise. In the mud crab *Panopeus herbstii* and the fiddler crab *Uca pugilator* the oxygen consumption under fluctuating temperature conditions in the 15°C to 25°C range was depressed relative to the acclimated rates of animals subjected to a constant temperature equivalent to the mean temperature in the fluctuating conditions (Dame and Vernberg, 1978).

To conclude, it is necessary to point out an important but sometimes ignored distinction between acclimation as defined above (adaptive adjustment observed after temperature manipulation) and the more complex alterations in metabolic rates observed in association with seasonal variations in environmental temperatures. These are often described as *acclimatisation.* They certainly include acclimatory adjustment but also involve metabolic shifts due to seasonal variations in behaviour patterns, growth rates and reproductive activity (Parry, 1978).

4 EXCRETION

4.1 What Is It?

Amino acids absorbed from the alimentary tract and in excess of metabolic requirements, together with those generated by the breakdown of body proteins, are broken down further in a process known as oxidative deamination:

$$\underset{\text{amino acid}}{NH_2\overset{\overset{\displaystyle R}{|}}{C}HCOOH} + 1/2O_2 \rightleftharpoons \underset{\text{keto acid}}{0 = \overset{\overset{\displaystyle R}{|}}{C}COOH} + \underset{\text{ammonia}}{NH_3}$$

The keto acids can be used in other pathways of intermediary metabolism. However, because of its low pH, ammonia is highly toxic to tissues and must be removed rapidly from the body or be rendered harmless. These processes are referred to as *excretion*.

Ammonia is the major waste product of protein and amino acid metabolism but the excess nitrogen need not leave the animal in that form – if, indeed, it leaves the body at all. As well as being highly toxic, ammonia is very soluble and is easily lost from aquatic organisms by a process of diffusion across the body or respiratory surfaces. Excretion which is dominated by ammonia is referred to as ammoniotelism. However, in terrestrial habitats this strategy would require the production of large quantities of urine and would, therefore, be expensive in terms of water loss, so here ammonia is usually converted to less soluble and less toxic products.

In terrestrial vertebrates urea is the predominant waste product and this strategy of excretion is referred to as ureotelism:

$$\text{urea} \quad \begin{matrix} NH_2 \\ | \\ C = O \\ | \\ NH_2 \end{matrix}$$

In the terrestrial invertebrates a variety of purines are used but uric acid production is most widespread. These strategies are referred to as purinotelism and uricotelism respectively:

purine (general structure)

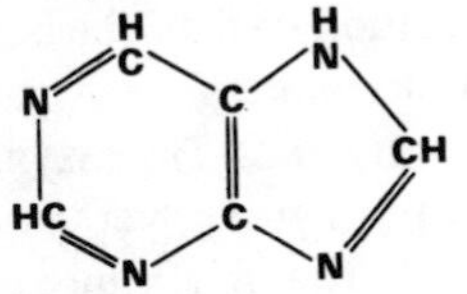

```
          H       H
          C       N
      N       C
                      CH
      HC      C
          N       N
```

uric acid

```
          O
          C       H
                  N
      HN      C
                      C=O
          C       C
      O       N       N
              H       H
```

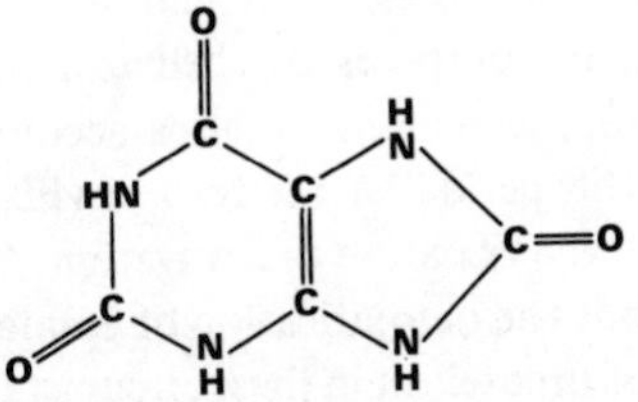

No one excretory product is used exclusively by any group of invertebrates. As already suggested the aquatic ones are predominantly ammoniotelic. Uricotelism is of importance in the Myriapoda and Insecta and to a lesser extent in the Crustacea and Mollusca. Only a few Platyhelminthes, Annelida and Mollusca have been found to be ureotelic. The main excretory product in the Arachnida is guanine, another purine:

guanine

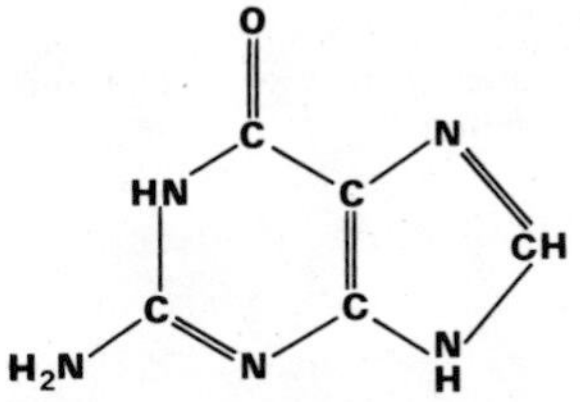

```
          O
          C       N
      HN      C
                      CH
          C       C
  H2N         N       N
                      H
```

A major exception to the general pattern outlined above is found in the Crustacea. Here ammoniotelism predominates, even in terrestrial species. In the latter, ammonia is expelled as a gas and tissues appear to be less sensitive to its toxic effects than those in other taxa. Terrestrial crustaceans do not possess epicuticular wax which is present in insects and which when present effectively reduces diffusion. In the isopod crustaceans the release of gaseous ammonia can account for all the nitrogen being excreted.

Direct loss of ammonia from the surface of the tissues not only serves the excretory needs of crustaceans but also assists in the formation of their exoskeleton. In particular the alkaline conditions created at the surface by this process provides CO_3^{2-} (from dissolved CO_2 as bicarbonate) for $CaCO_3$ precipitation:

$$NH_3 + HCO_3^- \rightarrow HN_4^+ + CO_3^{2-}$$

and this is used for hardening the exoskeleton. Molluscs may also make use of ammonia in the formation of their shells.

Another exception to the more general pattern outlined above is ureotelism in the pulmonate molluscs. The enzyme machinery to produce urea is widely distributed in this group, yet ureases often break the urea down as soon as it is formed. The significance of this is uncertain but it might be concerned with the transport of ammonia, in a harmless form, to the external surfaces for purposes of shell formation (see above). In *Bulimulus,* a small tropical pulmonate, urea accumulates in the tissues dramatically during the period of aestivation (Figure 4.1) only to be excreted rapidly upon the cessation of aestivation. Again the significance of this is uncertain, but the detoxification of ammonia is unlikely to be the sole importance of ureotelism in these circumstances. One possibility is that as the concentration of urea increases in the body fluids, the enhanced osmotic pressure will reduce evaporative water loss and this might increase survival during prolonged drought. In *Bulimulus* aestivation occurs during the dry season.

The low toxicity of purines means that they need not be excreted at all — and this seems to be what happens in a number of the shorter-lived Mollusca. For example Duerr (1966) put *Lymnaea stagnalis* (a

Figure 4.1: Accumulation of urea (μ mol.g^{-1} tissue) and uric acid (μ mol.g^{-1} tissue) in *Bulimulus* during aestivation.

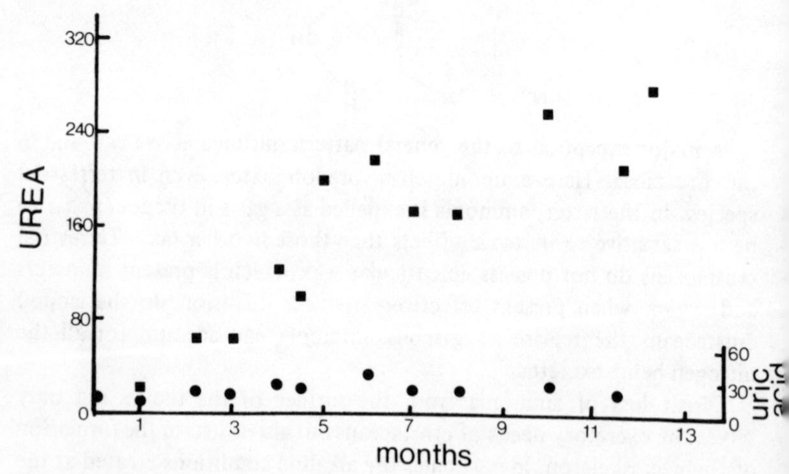

Source: Modified from Horne (1971).

freshwater pulmonate that lives for no longer than two years) in a solution of water and penicillin (to stop bacterial ammonia production) for 24 hours. After that time the solution was analysed for ammonia, urea and uric acid but none was found. Previous experiments, by other workers, had suggested that ammonia was being excreted, but no antibiotics were used and Duerr suggests that bacterial contamination was the most likely explanation of these results. On the other hand, uric acid accumulates in the tissues of this species as individuals get bigger and older, and the same phenomenon has been observed in several other aquatic snails (Figure 4.2). Badman (1971) has reported uric acid and guanine accumulation in the tissues of the small terrestrial pulmonate, *Mesomphix vulgatus,* but no loss of excreta. He points out that guanine is a more efficient purine for nitorgen storage than uric acid since it carries more nitrogen per unit mass of carbon.

Uric acid storage also occurs during the development of organisms in cleidoic eggs (i.e. those with membranes or shells which box the embryo off from the outside world), e.g. as shown in Table 4.1. Indeed the evolution of the cleidoic condition may have been a prime impetus for

Figure 4.2: The relationship between uric acid content ($\log_e$ mg/g) and snail dry weight ($\log_e$ mg). Each line represents a separate species.

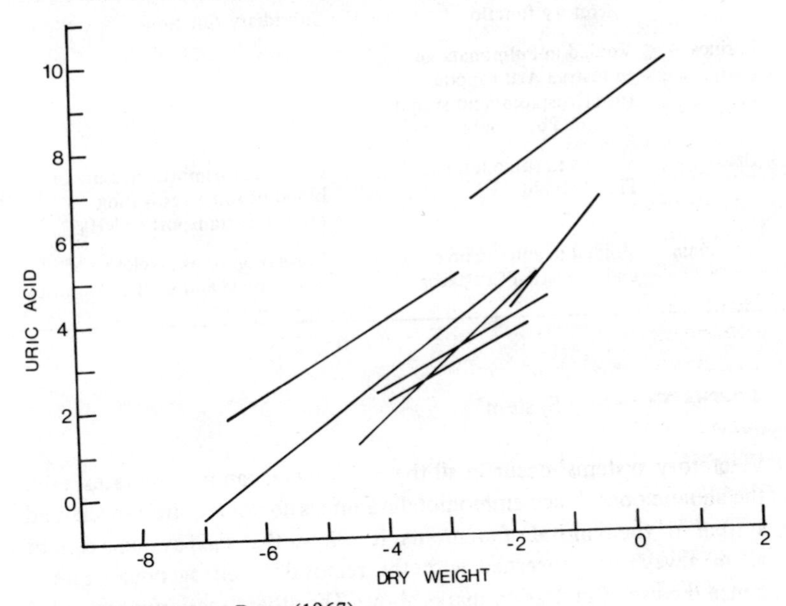

Source: Modified from Duerr (1967).

Table 4.1: Accumulation of uric acid in the cleidoic eggs of *Lymnaea stagnalis.*

Stage	Mg uric acid %
Cleavage	0.5
Foot and shell form, embryo occupies <¼ of the egg case	0.8
Embryo occupies ¼ to ½ of egg case	1.0
Embryo occupies ½ to ¾ of egg case	2.4
About to hatch	4.5

Source: Data from Baldwin (1935).

the evolution of purinotelism and both are of fundamental importance for the conquest of land (Needham, 1938).

A summary of the diversity of excretory products found in the invertebrates and the diversity of functions to which they are put is given in Table 4.2.

Table 4.2: Functions of excretory products of invertebrates.

	Excretory function	Subsidiary function
Purines	Voided in Pulmonata and terrestrial Arthropoda (not Crustacea) and stored in some Pulmonata	
Urea	Voided in some terrestrial Platyhelminthes and Annelida	Control of osmotic pressure in blood of some aestivating Mollusca: transport of NH_3
Ammonia	Voided in aquatic invertebrates and terrestrial Crustacea	Formation of exoskeleton of Arthropoda and shell of Mollusca

4.2 The 'Excretory System'

'Excretory systems' occur in all the major metazoan invertebrates, even the aquatic ones. Since ammoniotelic animals do not require a specialised system for the removal of ammonia, it follows that such systems cannot all or always be concerned with the removal of nitrogenous waste – hence the use of quotation marks above. The other functions of so-called 'excretory systems' are concerned with ionic and osmotic regulation, i.e. pumping out excess water (particularly in freshwater animals) or the differential removal and retention of specific ions.

Two kinds of excretory system can be distinguished on the basis of embryological origins:

(1) Nephridia – of ectodermal origin; develop from the external surfaces and grow into the body; two major types:

(a) *Protonephridia* (Figure 4.3a, b) – consist of a *closed* system of tubules ending in flame cells (e.g. Platyhelminthes and Nemertea) or solenocytes (e.g. primitive Annelida).

(b) *Metanephridia* (Figure 4.3c) – System of tubules which *open* into the body cavity as a ciliated funnel (e.g. oligochaete Annelida).

(2) Coelomoducts (Figure 4.3d) – mesodermal out-growths of the gonadas; develop from the inside and grow out (e.g. coxal glands of Arachnida, renal organs of Crustacea and 'kidneys' of Mollusca).

A number of systems are combinations of the above; for example in some annelids the nephridia and coelomoducts coalesce to form a nephromixium (Figure 4.4). Not related to this system of classification in any obvious way are the longitudinal excretory canals and the 'ventral glands' of the Nemathelminthes (of uncertain origin; Figure 4.5) and the malpighian tubules of Insecta which are outgrowths of the gut and hence of endodermal origin (Figure 4.6). Echinoderms and cnidarians do not have excretory systems at all.

Metanephridia, excretory glands and the molluscan 'kidney' work by a system of ultrafiltration – the filtrate passing from a region of high pressure in the body fluids to the system itself. There is little modification in the composition of the fluid itself during this process but, once in the system, it may be modified by differential absorption and secretion. In most terrestrial animals, for example, water is resorbed. Modification of the ionic composition of the urine of the crayfish as it passes through the renal organ is illustrated in Figure 4.7.

How protonephridia work is not fully understood, but one possibility is that the movement of cilia in the closed tubules creates sufficient negative pressure for ultrafiltration. In malpighian tubules, however, urine is produced by a process of secretion rather than ultrafiltration, i.e. both water and ions are secreted into the ducts of this system. Most of the water lost in this way is resorbed by the rectal gland (Figure 4.6) before leaving the alimentary canal. Table 4.3 shows the relative compositions of serum, urine and rectal fluid of the stick insect.

Figure 4.3: Nephridia and coelomoducts in a hypothetical coelomate 'worm'. a = protonephridium with flame cell (contains a number of flagellae, the flame); b = protonephridium with solenocyte (contains single flagellum); c = metanephridium; d = coelomoduct. N.B. No single invertebrate has all these structures.

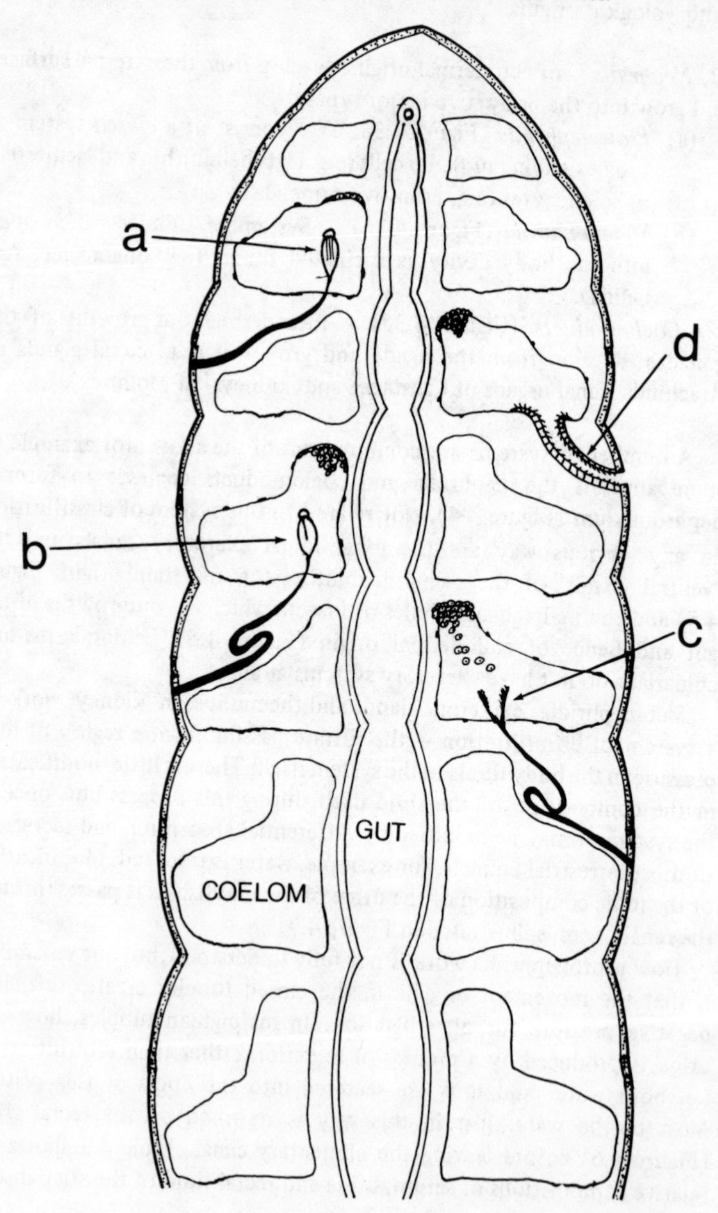

Figure 4.4: Nephromixium. M = mesodermal (gonadal) component; E = ectodermal (nephridial) component; G = gamete.

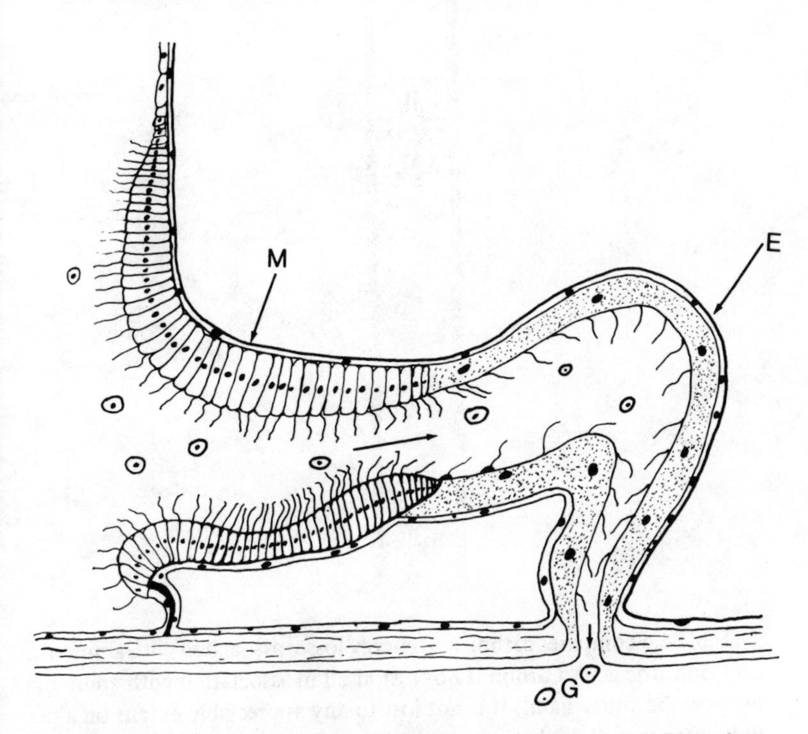

In this animal potassium becomes concentrated in the urine. Potassium urates are also secreted into the malpighian tubules. These are kept in solution at the relatively high pH at the distal part of the tubule but are precipitated either at the end of the malpighian tubule or in the rectal gland as water is resorbed and the pH rises.

4.3 Energy Costs and Benefits

The benefits of purinotelism derive from the low toxicity of the purine products. Needham (1931) suggested that costs might also derive from this strategy since being more complex than the other excretory products, purines will lead to the loss of more carbon and potential energy in association with the excess nitrogen. Ureotelism stands somewhere between ammoniotelism and purinotelism in this respect.

Figure 4.5: An H-shaped excretory system of a nematode. There are two ventral gland cells.

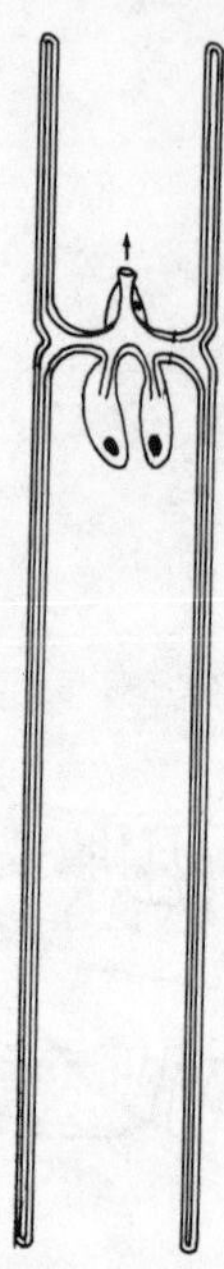

Table 4.4 compares carbon and energy losses associated with ammonia, urea and uric acid. Carbon is not lost at all in association with ammonia but, on the other hand, it is not lost to any appreciable extent on a per unit nitrogen (C/N) basis in association with urea and uric acid. Also, though the energy loss per mole of product is much greater for uric acid than urea, and for urea than ammonia, the differences turn out not to be significant when calculated on a per unit nitrogen basis. Indeed when calculated in this way the energy loss associated with urea is less than that associated with ammonia. The energy used in the formation of these products might be considered more important and urea and uric acid certainly do use more energy in this respect (see the final two columns of Table 4.4).

Another possible argument is that the more complex excretory products require more complex apparatus to process them. These, in turn, would require more energy and resources to build and maintain them. However, against this, it is to be remembered that 'excretory systems' occur whether or not they are required for excretion and there

Figure 4.6: Insect excretory system. M = malpighian tubules; RG = rectal gland; – food; - - - water; . . . uric acid.

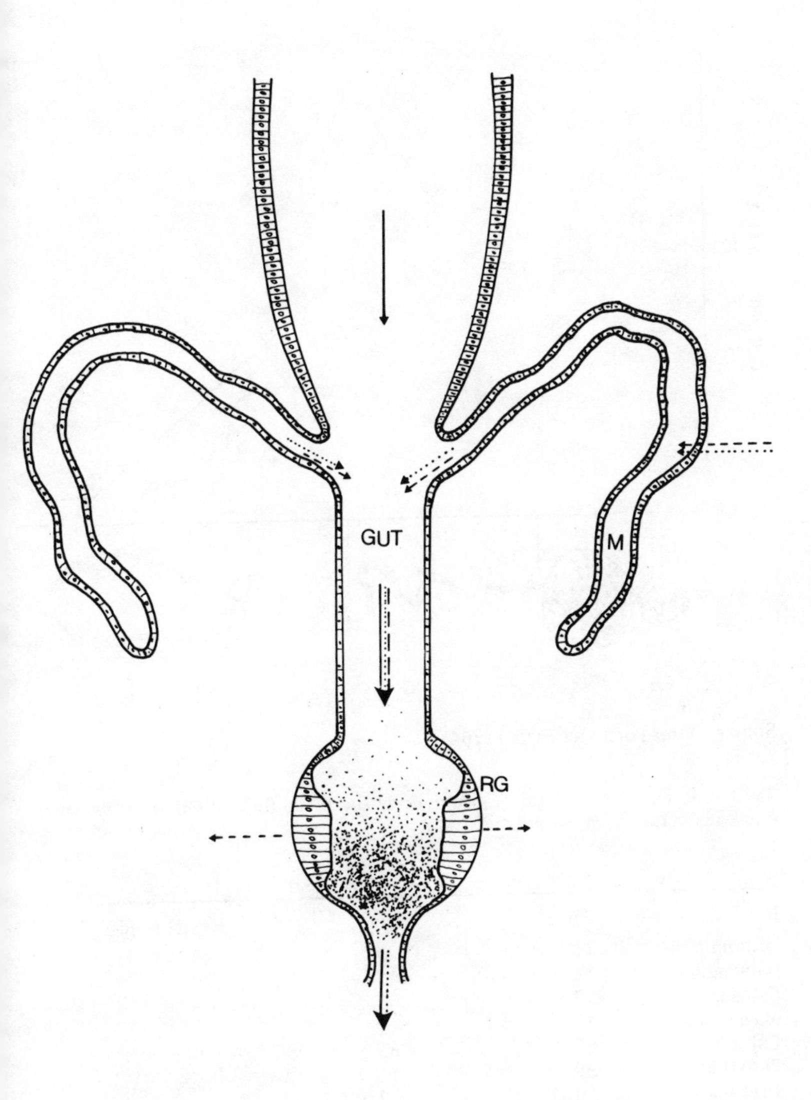

Source: After Potts and Parry (1964).

Figure 4.7: Change in chloride composition of urine as it passes along the renal organ of a crayfish. ES = end sac; L = labyrinth; NC = nephridial canal; B = Bladder.

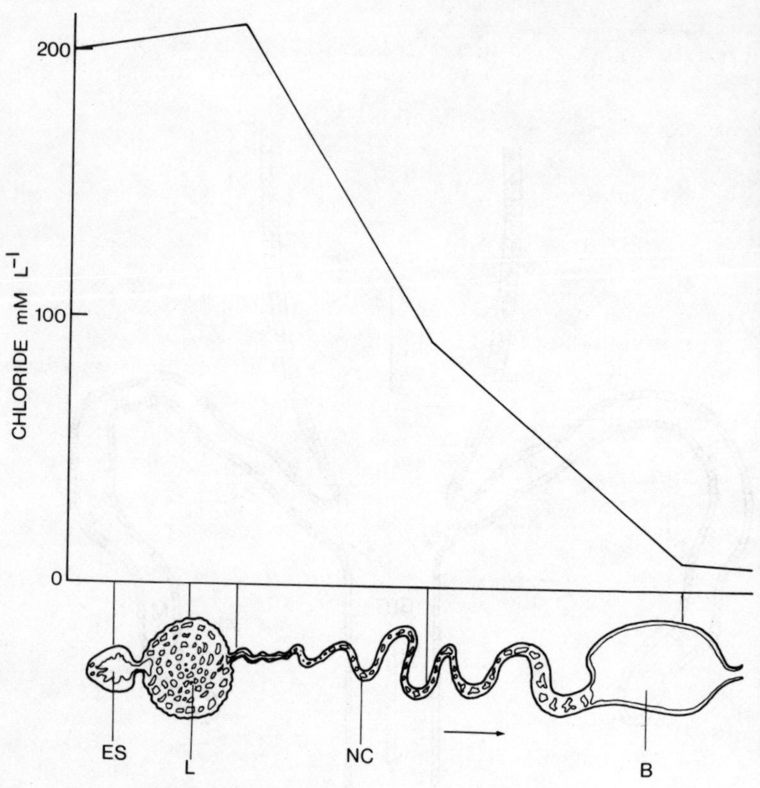

Source: After Potts and Parry (1964).

Table 4.3: The composition of serum, urine and rectal fluid of *Carausius morosus* (stick insect).

Ion	Serum	Urine	Rectal fluid
Sodium	11	5	19
Potassium	18	145	320
Calcium	3.5	1	
Magnesium	54	9	
Chloride	87	65	
Phosphate	39	51	
Uric acid	0.27	2.6	
pH	6.6	6.8-7.5	3.5-4.5

Source: Modified from Potts and Parry (1964).

Table 4.4: Carbon and energy costs in nitrogen excretion.

	C/N	Heat of combustion		Free energy of formation	
		$kJ.mole^{-1}$	$kJ.mole^{-1}.N^{-1}$	$kJ.mole^{-1}$	$kJ.mole^{-1}.N^{-1}$
Ammonia	0	378	378	27	27
Urea	0.5	638	319	205	102
Uric acid	1.25	1932	483	384	96

Source: Data from Pilgrim (1954).

appears to be no obvious correlation between the complexity of 'excretory systems' and the products they manufacture.

To summarise then: urea and uric acid are more expensive in carbon and energy loss than ammonia, but the differences are not as substantial as might have been expected. Hence, economies in water loss and relative toxicities are likely to have been more potent forces in the evolution of excretory strategy than the extent of energy and carbon losses.

4.4 Secretions

As much if not more energy and resources may be lost from the body of invertebrates in secretions as in excretions. Mucous losses may amount to 20 per cent of the total energy flow in snails and more in triclad Platyhelminthes. Of course, mucous loss is not all cost. Locomotory benefits accrue in snails and triclads which use mucous trails as a lubricant and to give some point of leverage for cilia. Mucous trails may also be used as a trap for food. These matters were discussed more fully in Chapter 2.

5 GROWTH

5.1 Introduction

This chapter is about growth, an obvious process but one which has proved difficult to define precisely and rigorously. Here, it is simply used to mean increases in the sizes of whole organisms with time. Size, of course, can be expressed either in linear dimensions, mass or potential energy and this in itself has presented one of the major difficulties for formulating a precise definition – which is the most appropriate unit? Some consideration will be given to the problem of mensuration in Section 5.3. Most of the chapter will be concerned with a comparative approach to the mechanistic basis of increases in size and, as usual, to their adaptive significance. Ageing phenomena will also be discussed at the end of the chapter since the initiation of ageing has, for many years, been associated in the minds of biologists with the cessation of growth.

5.2 Metabolic Basis

Growth occurs when the input of nutrients to the organism is more than enough to make good the losses incurred in the dissipation processes. In other words it is a product of metabolism. The pattern of size-change with time depends upon how size itself influences the gains and losses of resources by the organism. Bertalanffy (1960) modelled growth (in mass) as a balance between anabolic (building-up) and catabolic (dissipation) processes using a very simple equation:

$$dM/dt = \zeta M^n - \chi M^m \tag{5.1}$$

where: M = body size; t = time; ζ, χ, m, n are constants. Winberg (1956) used a similar approach but replaced the abstract concepts of 'anabolism' and 'catabolism' with resource inputs and outputs and made his model self-consistent by expressing all terms, input, output and size, in energy units:

$$\Delta W/\Delta t = C - R \tag{5.2}$$

where: W = energy equivalent of size; t = time and $\Delta t \gg dt$; C = energy acquired in food; R = energy expended in metabolism. Energy input can be measured as food ingested (I) or absorbed (A) across the gut wall and energy output as excretion and respiratory heat loss. In a wide variety of animals we have already established that respiratory losses increase as body mass (M) increases but at a reducing rate (Chapter 3) such that:

$$R = aM^b \tag{5.3}$$

Less is known of the relationship between food input (C) and body size but a similar kind of relationship as expressed in equation 5.3 with the exponent tending to 0.67 (i.e. intake is surface-dependent) has been found for Anthozoa, triclad Platyhelminthes and some Gastropoda (Calow, 1981). Therefore:

$$C = cM^d \tag{5.4}$$

Taking logarithms of equations 5.3 and 5.4 yields equations for straight lines ($\log R = \log a + b (\log M)$ and $\log C = \log c + d (\log M)$, and a number of growth types can be defined from the relationship between these two lines (Figure 5.1).

Assuming that the rate of input begins at a higher level than the rate of output ($a < c$), then the difference between the two lines represents energy (E) which is available for the accumulation of biomass or in other words for growth. If $b > d$ the input and output lines must ultimately intersect, i.e. where $E = O$. At this point no further increase in size is possible and so growth is said to be *limited* (Figure 5.1a). The way size approaches the limit (shape of the growth curve) depends upon the way the input and output lines alter with size. Often the growth curve is sigmoid, with size decelerating onto a steady-state after an initial period of exponential growth. Typical examples are shown in Figure 5.2.

When $b = d$, growth is *unlimited* (Figure 5.1b) and the accumulation of biomass will proceed at an exponential rate. Non-metabolic (e.g. mechanical) limits may then intervene when a sharp cut-off in growth is expected and 'J' shaped growth curves will result. Apparently unlimited growth also arises from metabolic systems in which the input and output trajectories intersect if this happens at a point when the organism is normally dead (Figure 5.1c).

Finally, it is worth distinguishing *growth limitation* from *growth regulation*. Environmental variations, for example in food supply and

Figure 5.1: Functional dependency of energy input (C) and output (R) on the size (M) of animals. The skull and cross-bones marks the point of death and the star the steady-state size. E is the difference between C and R and is the energy available for growth.

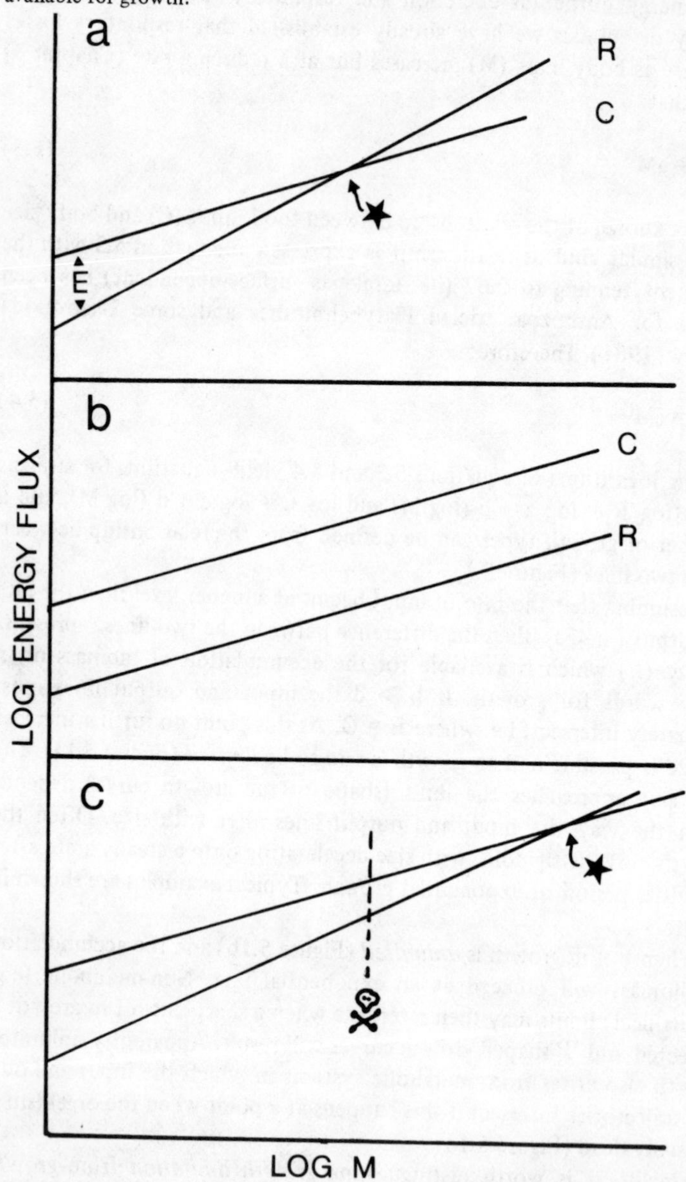

Source: After Calow (1981).

Figure 5.2: Sigmoid growth curves of several invertebrates: (a) shell diameter (mm) of a planorbid snail; (b) body plan area (mm^2) of a rhabdocoel flatworm; (c) joules potential energy of an erpobdellid leech; (d) mg fresh weight of a slug. Each sigmoid curve consists of an accelerating (exponential) part and a decelerating part on to a final, adult size. There is, therefore, an inflexion in the curve marking the point of transition between the two parts. The accelerating part can be explained as a result of growth processes producing tissues capable of more growth. The decelerating part is due to the constraints of size imposing themselves on this process.

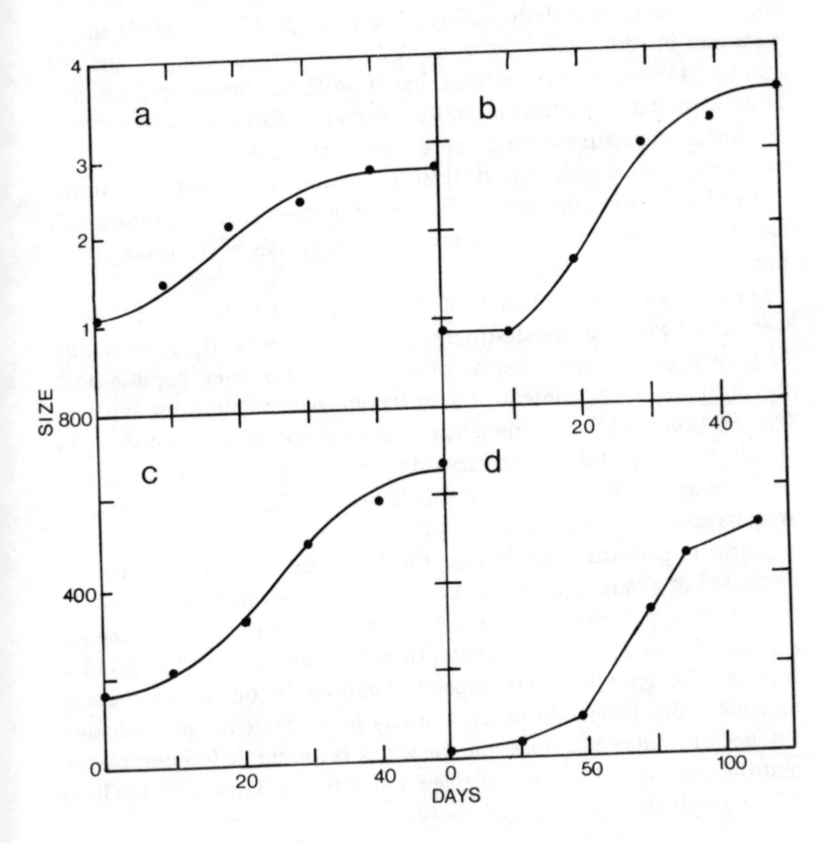

temperature, are likely to alter the metabolic parameters of the input and output processes and thereby shift the limits (if there are any) to growth and alter growth rates. However, there may be some metabolic resistance to these shifts such that compensating alterations in metabolism, relative to disturbance, keep growth rates and final size on pre-(perhaps genetically) determined targets. This is referred to as *regulated* growth.

5.3 Distribution of Limited and Unlimited Growth

5.3.1 Units

No one measure of size is completely satisfactory. Potential energy is useful in that it gives a good index of metabolising biomass and in that food inputs and respiratory and excretory outputs can be measured in the same units. However, though the energy contents of tissues can easily be determined, e.g. by bomb-calorimetry (Figure 2.1); this can only be carried out destructively. The same problem arises if size is measured in terms of dry weight, carbon or nitrogen content and cell number. Hence, growth patterns, based on these parameters, can only be constructed by plotting mean sizes from population data against time and these procedures often obscure important features of the individual patterns. For example, the differential mortality of size-groups might cause alterations in the pattern of changes in mean size with time which are quite unrelated to size changes in the individuals which make up the population.

In contrast, the estimation of size in terms of linear dimensions or wet (fresh) weight is non-destructive and can be used in the construction of growth curves representative of individuals. However, because both the form and water content of animals can change during development (e.g. Section 5.8) these parameters again need not be related directly or simply to the quantity of metabolising biomass in the animal. Furthermore, in soft-bodied invertebrates it is often difficult to estimate fresh weights accurately and consistently.

Growth patterns described in the literature for invertebrates are expressed in a variety of ways and convenience seems to have played a predominant part in determining which parameter has been chosen for a particular study. It is fortunate, therefore, that though the detailed form of the growth curves depends importantly on the way size is measured, the general form, particularly in terms of the presence and absence of limitations, does not. Hence, it is legitimate to interpret the multifarious facts that are available on invertebrate growth on the basis of the simple theory developed above.

5.3.2 Growth of Invertebrates with No Exoskeleton

Unlimited, exponential growth is only possible when weight-specific nutrient intake can keep pace with weight-specific respiratory losses. Exponential growth with a sharp cut-off has been recorded for many cnidarians (Figure 5.3) and here both food intake and respiratory losses are likely to be surface-dependent (Calow, 1981).

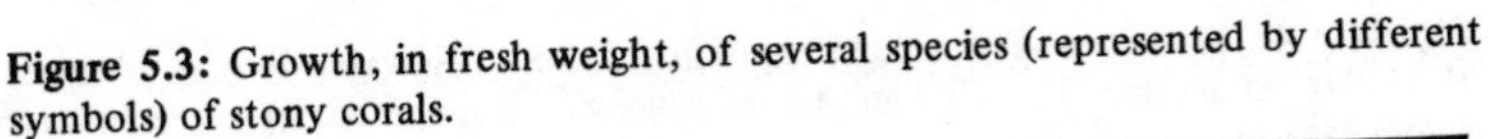

Figure 5.3: Growth, in fresh weight, of several species (represented by different symbols) of stony corals.

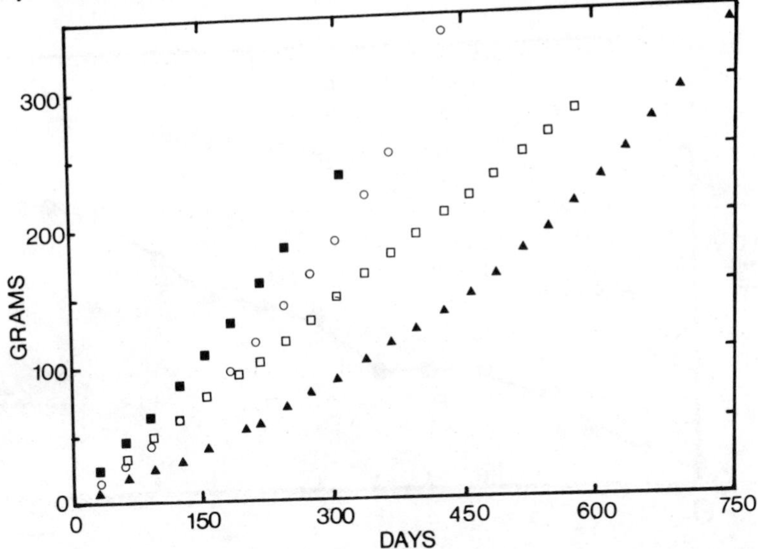

Source: Modified from Bak (1976) after Calow (1981).

A few other groups are known to grow in this exponential fashion under the appropriate circumstances (e.g. some flukes, tapeworms and bivalves) but amongst most invertebrates, sigmoid growth predominates in laboratory-reared individuals, viz: free-living Platyhelminthes, Nemathelminthes, Annelida and Mollusca (Figure 5.2). In the latter case, though, size-plateaus may not be reached before animals die (Figure 5.1c) and this is particularly true of the Cephalopoda.

Under natural conditions, of course, the sigmoid growth pattern often becomes more complex. This is because variations in temperature and food supply cause fluctuations in growth rate and may lead to temporary cessations and 'spurts'. Figure 5.4 shows a complex, triple-sigmoid growth pattern in individuals from a field population of the fresh-water limpet, *Ancylus fluviatilis.* Under laboratory conditions this species has a simpler, sigmoid, growth pattern.

5.3.3 Growth of Invertebrates with an Exoskeleton

Bertalanffy (1960) claimed that the weight exponents of the anabolic and catabolic processes in the Insecta are equivalent and tend to one. Accordingly growth should be of the accelerating, unlimited kind and

Figure 5.4: Shell-free dry weight growth of the freshwater limpet, *Ancylus fluviatilis,* in nature. The first month is August – i.e. summer. Data points are population averages.

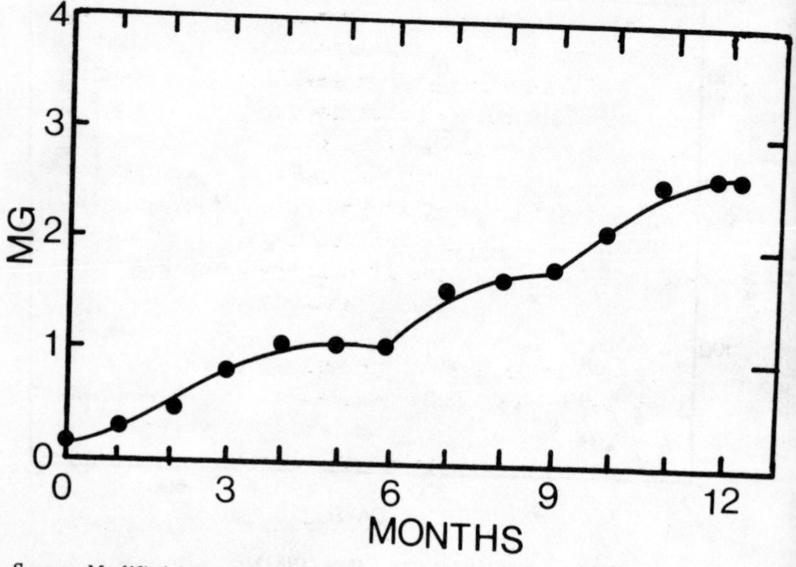

Source: Modified from Calow (1981).

this is indeed true for some species (e.g. as shown in Figure 5.5). Nevertheless insects have exoskeletons and these introduce some complications.

The insect cuticle is like a rigid box which prevents free expansion, by growth, of the internal organs. Therefore as insects increase in size their exoskeleton must be shed and replaced – a process referred to as moulting or ecdysis. For a few hours after the old cuticle is shed the new one is pale and soft. During this period the insect enlarges to the size of the next stage usually by taking in air in terrestrial species or water in aquatic ones (Figure 5.5b). Hence the growth in length of insects, or in the fresh weight of aquatic species, occurs in a step-wise fashion. Dry weight increases more continuously between moults replacing the air or water. However, there may be a slight reduction in the fresh weight of terrestrial species at each ecdysis (Figure 5.5a) due to the loss of cuticle and to the loss of water which is not replaced because the insect does not feed or drink during the moult. The blood-sucking bugs (e.g. *Cimex* and *Rhodnius*) have an even more curious growth curve. They take a single, large meal during each moulting stage which is metabolised between moults, so the weight curve shows a

Figure 5.5: Growth patterns in insects: (a) a terrestrial species (e.g. *Dixippus*); (b) a freshwater species (e.g. *Notonecta*); (c) a blood-sucking species (e.g. *Rhodnius*). ▲= ecdysis; ★= final size.

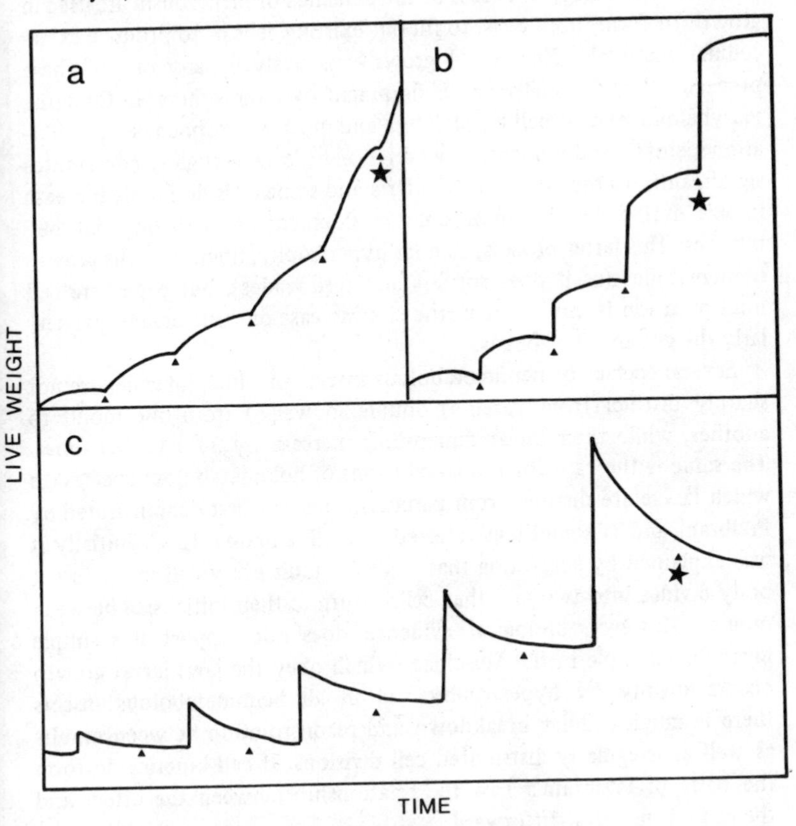

Source: Modified from Wigglesworth (1974).

series of acute peaks interpolated upon a gradual increase to the final size after the terminal moult (Figure 5.5c).

The exoskeleton puts limits on the final size which insects can achieve. Hence growth ceases abruptly after the last moult leading to 'J'-shaped growth curves.

Growth in other arthropods resembles this pattern observed in the insects. However, in nematodes, which also have a rigid cuticle, there is some growth of the cuticle between moults so that the step-wise increase in linear dimensions is not as pronounced.

5.4 Cellular Basis

Materials and energy in excess of the demands of metabolism are used in growth to form more cells, to fill out existing cells or to produce extracellular materials. No animal grows exclusively by any one of these processes though usually one is dominant over the others. In Cnidaria, Platyhelminthes, Annelida, Mollusca and most Arthropoda cell proliferation seems to be dominant. However, non-cellular mesogloea contributes significantly to the growth of Cnidaria and some cells definitely increase in size during the post-embryonic development of free living Platyhelminthes. This latter process, cellular hypertrophy, dominates the growth of nematodes (as it does rotifers and tardigrades), but even here cell multiplication is important in the size increase of some organs, particularly the gut and the gonads.

Several species of hemimetabolous insect (in which juveniles are not sharply distinct from parents) double in weight from one moult to another, while their linear dimensions increase by $3\sqrt{2}$ (1.26) times. The same is the case for the larval forms of holometabolous species (in which larvae are distinct from parents). This was first demonstrated by Przibram and is sometimes referred to as 'Przibram's Law'. Initially it was explained by suggesting that at each moult every cell in the insect body divides into two and that cells return to their initial size between moults. However, biological evidence does not support this simple idea; for example in the Muscidae (which obey the law) larval growth occurs mainly by hypertrophy and in all hemimetabolous insects there is much cellular breakdown and reconstruction between moults as well as irregularly distributed cell divisions. If cell kinetics do form the basis of Przibram's Law the relationship between the effect and the cause is not straightforward.

Cell proliferation often persists in non-growing, adult invertebrates when it must balance cell loss. This has been particularly well studied in *Hydra* where mitosis occurs throughout life and has been observed in all parts of the body. Here cell production is balanced by the loss of cells in asexual buds and also by some sloughing from the surface. Cell turnover probably occurs to a greater or lesser extent in the adults of all invertebrates but is extremely limited in extent and distribution in some groups. Growth may occur by cell proliferation in insects but in the adults most cells are in a post-mitotic state so that there is little cell turnover in these animals. What cell turnover there is, and this also applies to nematodes, is limited to the gut and gonads.

5.5 Adaptational Aspects

It was assumed, at the outset (Chapter 1), that selection would tend to maximise the energy available for growth (E) because, other things being equal, this should shorten the time between birth and reproduction and shorten the time spent by animals in the small, usually vulnerable stages of their life-cycles. Even if the breeding season is fixed and finite, a maximised growth rate will be favoured if it results in big animals and if, as is usual (Chapter 6), there is a positive correlation between fecundity and the size of the parent. Fitness, in other words, is positively correlated with survivorship and fecundity and negatively correlated with generation time.

Now growth rate is a function of both the speed of the metabolic, conversion processes and their efficiency. We might expect one or other or perhaps even both of these properties to be maximised. However, options are limited by the fact that the latter seems to be prohibited, possibly by physical constraints (Odum and Pinkerton, 1955).

The conversion process itself involves the synthesis of polymers from monomeric precursors derived from the food. This is an energy-consuming process which is powered by energy derived mainly from ATP. Hence, as is apparent from Figure 3.1, the overall efficiency of conversion depends crucially on the ATP-generating reactions. We have already considered these in Chapter 3. They are efficient but not, perhaps, as efficient as might be expected and the reason for this is most plausibly associated with constraints imposed by initial conditions (at the origin of life) and the requirements for flexibility in the reaction systems.

Given these constraints, and making certain minimum estimates about the demands of physical and chemical work, it is possible to calculate a best possible efficiency for overall conversion, and this works out at approximately 80 per cent (Calow, 1977a). Table 5.1 compares actual efficiencies observed in invertebrates with this theoretical limit. Note that the ranges that have been recorded sometimes stretch beyond the theoretical limit and that most fall well below the expected maximum. The higher efficiencies are often associated with sedentary animals (e.g. anemones and *Lestes sponsa,* a sit-and-wait predator) and this suggests that economies might be achieved by immobility. *Calanus,* a mobile and active planktonic copepod, also has a high efficiency. This may be because in the plankton it is often surrounded by phytoplankton, its source of nutrient, and does not incur heavy costs in catching and processing food. Overt activity is obviously expensive in energy but it must be remembered that it might still lead

Table 5.1: Conversion efficiencies (%) in a variety of invertebrates as compared with the theoretical maximum.

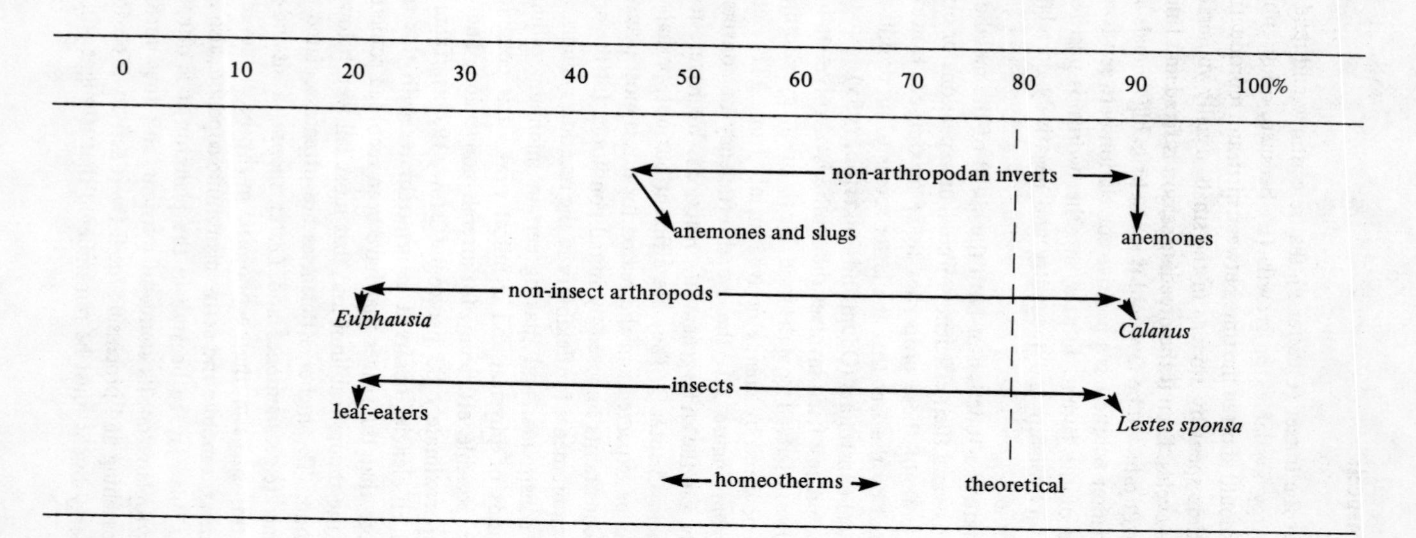

to compensating and, possibly, overriding fitness gains – for example in escaping a predator or in capturing prey. Indeed the ability to move has been a major requirement in the evolution of most heterotrophs.

Also included in Table 5.1 is the range of efficiencies for warm-blooded vertebrates (homeotherms). This range does not reach the theoretical maximum, presumably because of the cost of endothermy – generation of heat to maintain a constant and above-ambient body temperature. The gains from such a strategy derive from the possibility of high and constant rates of conversion despite low and fluctuating environmental temperatures. This is illustrated in Figure 5.6 which summarises, as regression lines, the relationship between the growth rates ($\log_{10}$ g/day) and adult sizes ($\log_{10}$ g) of several species of vertebrates. These are extracted from a review published by Case (1978). Though possibly having lower conversion efficiencies, on a size-for-size basis, mammals had higher growth rates than reptiles and fishes. Data-points for invertebrates (all except the parasite corrected to 10°C) are superimposed on these lines and conform more to the cold- than the warm-blooded vertebrates. There are a number of interesting points to be noted:

(1) *Hymenolepis diminuta* has high growth rates, tending to those of the homeotherms. This is a parasite of rodents and may be benefiting from the homeothermic properties of its hosts. The fact that parasites are surrounded by a superabundant supply of food may also enable them to be efficient converters.
(2) The octopus has a very rapid growth rate and this may be associated with its sit-and-wait feeding strategy; i.e. it is probably a very efficient converter.
(3) When the invertebrate data are corrected to 38°C – which is a typical body temperature for homeotherms – growth rates increase and are then as good as, if not better than, those for the mammals. These corrected data are represented by the broken line in the figure. This result suggests: (a) that despite acclimatory phenomena invertebrate growth rates are constrained in nature by temperature (in the temperate world 10°C is a typical average temperature); (b) that given equivalent conditions, invertebrates ought to do as well as, if not better than mammals – presumably because they are more efficient converters. This is not to say that invertebrates would actually grow better than mammals at 38°C because their enzymatic machinery is not adapated for operation at these high temperatures. Denaturation rather than rate-effects would therefore be likely

Figure 5.6: The relationship between $\log_{10}$ growth rate in fresh weight and $\log_{10}$ adult size of a variety of animals. Line m is for mammals, r for reptiles, f for fishes. Key to invertebrate species: P = Platyhelminthes; H = *Hymenolepis diminuta;* N = Nemathelminthes; O = Oligochaeta; L = Hirudinea; M = Mollusca; Ot = *Octopus;* E = Echinodermata; I = Insecta; C = Crustacea; A = Arachnida. Broken line is for invertebrate data corrected to body temperatures typical of mammals.

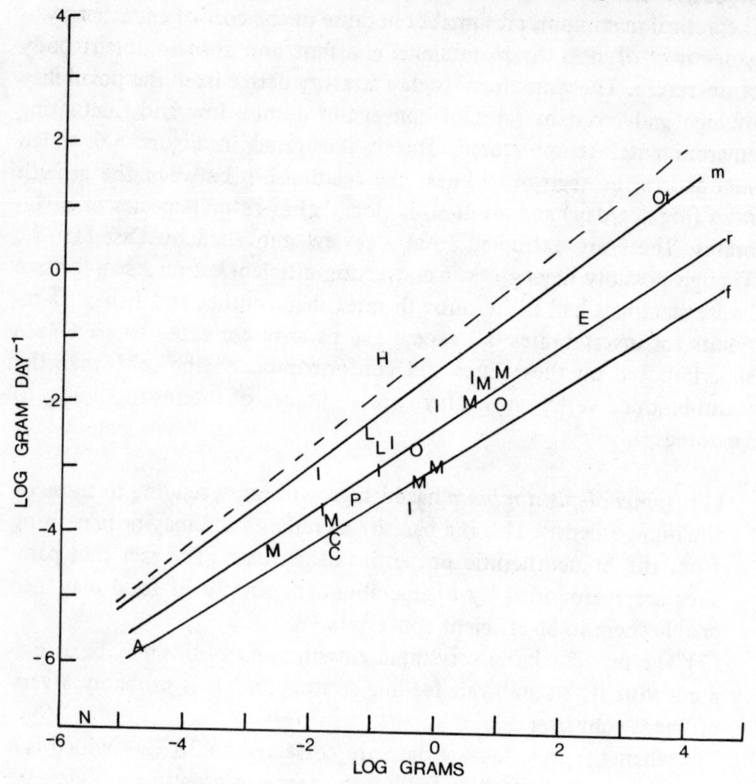

Source: Partly from Case (1978), with material modified from Calow and Townsend (1981).

to dominate under these conditions.

Clearly, then, if growth rates are maximised they are so within the biochemical, physiological and presumably developmental constraints associated with a particular organisation. But does selection actually 'push' growth rates to these theoretical limits? One set of observations suggests not. In some invertebrates, growth rates increase above normal when conditions ameliorate after a period of poor growth, e.g. because of starvation or a challenge from some toxins (Figure 5.7). These 'spurts' are thought to be compensating 'attempts' to return the organism to its normal size for age as soon as possible. That they occur suggests that under normal circumstances the growth rate of the organism is actively regulated below the maximum possible level, perhaps by inhibitory factors. This might be the manifestation of yet another developmental constraint – for example, putting the organism together too rapidly might lead to developmental errors. Or, they might be due to ecological and behavioural constraints – for example, an organism which grows more rapidly than others might become sufficiently different from the rest to attract the attention of predators. What particular constraints are at work will vary from species to species and population to population but the main point is that rather than being maximised in the true sense of the word, growth seems just as likely to be optimised.

Figure 5.7: Growth rates of two species of planarian after various periods of starvation. Growth rate at time zero is the control. Coefficient of exponential growth is the measure of the rate of logarithmic increase over the initial, exponential phase of growth. Surprisingly, the growth rates of these animals were not impaired even after two months continuous starvation! Over the first month there was some enhancement of growth – i.e. a growth spurt – for at least one of the species.

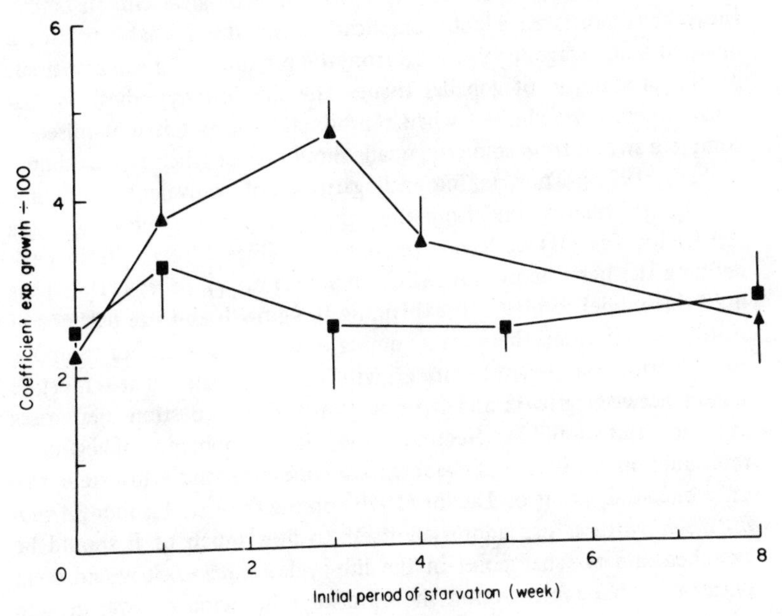

Source: With permission from Calow and Woollhead (1977).

The implication of this conclusion for the maximisation principles that have been applied in Chapters 2 and 3 to nutritive income and respiratory expenditure is that such principles must be tempered by the *active regulation* associated with optimisation. One model, showing how a control system might be grafted on to the energy budget of an animal, is illustrated in Figure 5.8. An explanation of the model is given in the legend since the reader will best follow it by reading this in close conjunction with the figure. What is eaten, and how much, is influenced by appetite which might, in turn, be influenced by the growth program and how well the organism is on target. Similarly the power output of the respiratory system and how the ATP is used might also depend on the state of the system relative to what is specified by the developmental program.

5.6 On When To Stop Growing

Returning to Figure 5.1, we have seen that growth ceases when the input and output 'lines' intersect, i.e. E = O. Selection should have produced a growth sequence and final size which, if not maximised, will be optimised with respect to the prevailing ecological circumstances. However, there is an added complication and this is that a part of E must, at some stage, be switched from the production of somatic tissue to the production of gonadal tissue. The simplest hypothesis is that since fitness is maximised when reproductive output is maximised, a complete switch from soma to gonads should occur when E is maximum (Sebens, 1979). Thus, taking antilogarithms of equivalent inputs and outputs, subtracting one from the other and plotting the differences (E) against size (M) we have a peaked curve (Figure 5.9) with the peak defining the best theoretical switching point (M_{opt}). Sebens (1979) has used this model to effect in explaining the growth and size patterns of Anthozoa. However, there are a number of complications. For example, reproduction may begin before growth has ceased such that E must be shared between growth and reproduction and the question then arises as to how this should be effected. Though the phenomenon of beginning reproduction before growth has ceased is widespread in the invertebrates, only one study, that of Lawlor (1979) on the terrestrial isopod, *Armadillidium vulgare,* has addressed itself to how much of E should be switched and at what point in the life-cycle. *A priori,* it would seem reasonable that reproduction should always be favoured over growth unless the products of reproduction stand a poor chance of surviving to

Figure 5.8: A cybernetic model of growth. Actual growth rate is represented as the summation of food inputs (A) and respiratory outputs (R). This, summed or integrated with time (∫ ( ) dt), gives current size which has a positive influence on A and R since both are size-dependent. A growth program (P), perhaps genetically determined, specifies a target growth rate. The target is compared with the actual by subtracting one from the other and the difference is an error-correcting signal (i.e. which is zero with no misalignment; is positive if the target is greater than the actual; and is negative when the actual response overshoots the target). The error-correcting signal feeds back negatively on R (i.e. turning it down if actual growth is poor) and positively on A (i.e. increasing the tendency to feed if growth is poor). D = environmental disturbances. Note that this is only one possible idea on how metabolism might be controlled. For others see Source.

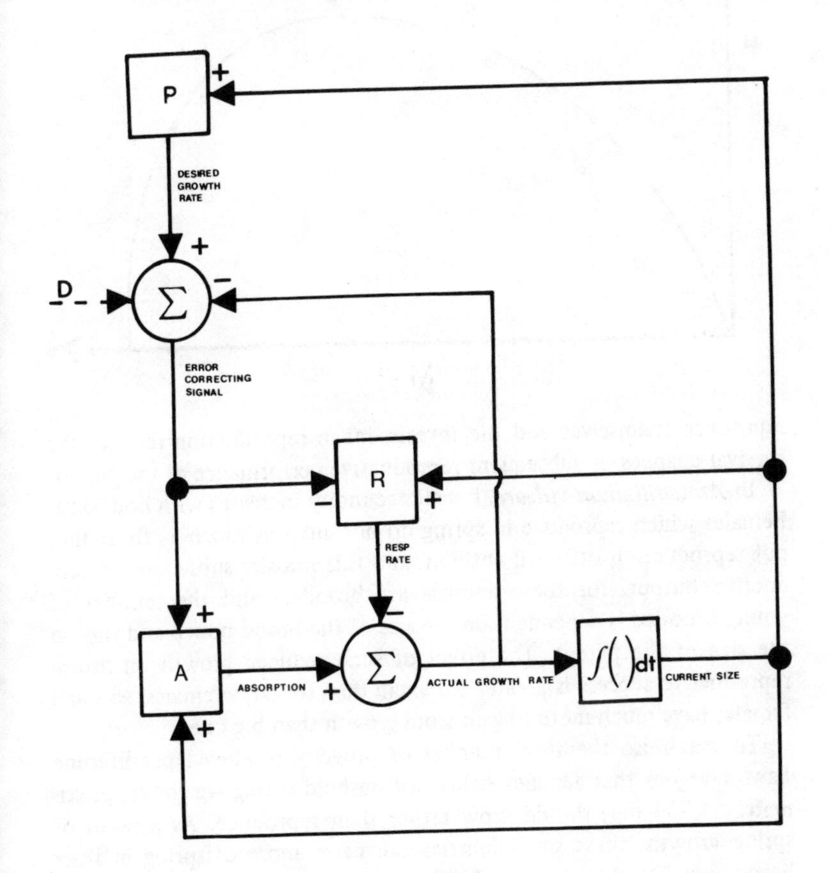

Source: Modified from Calow (1976).

Figure 5.9: A plot of E versus M in a sigmoid grower.

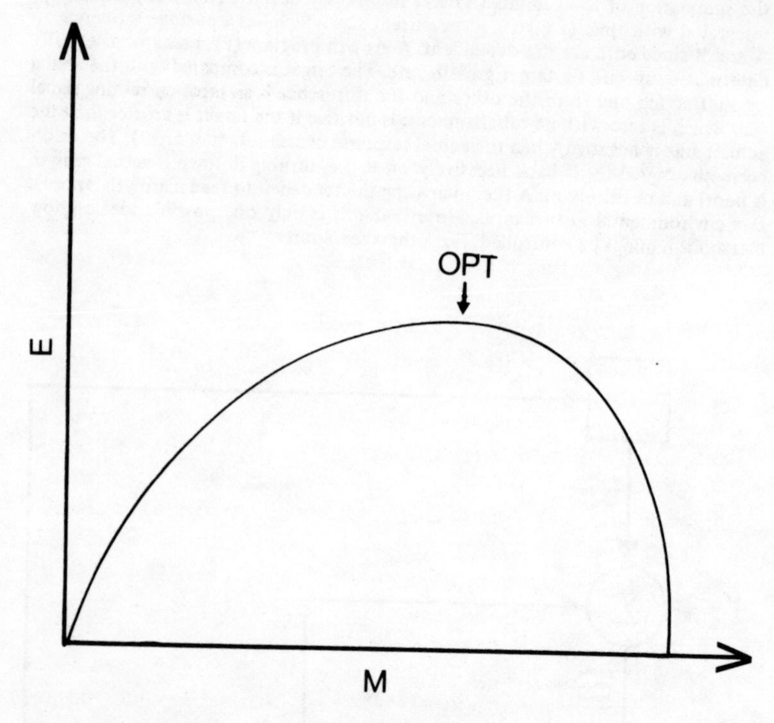

reproduce themselves and the investment in reproduction reduces the survival chances or subsequent reproductive performance of the parent.

In *Armadillidium vulgare,* female fecundity increases with body size. Females which reproduce in spring do not grow as much as those that put reproduction off until autumn, and this impairs subsequent reproductive output, for these animals are brooders and the number of young brooded is dependent on the size of the brood pouch and this on the size of the parent. The effect of such reduced growth on future reproductive success is greater for small than for large females, so small females have much more to gain from growth than big ones.

To maximise the total number of progeny produced per lifetime Lawlor argues that females below a threshold spring weight (approximately 45-50 mg) should grow rather than reproduce. As a result of spring growth, these small females can carry more offspring in their brood pouches than they could if they opted for reduced growth and both a spring and an autumn brood (i.e. the sum of two such small broods is less than one single large brood). However, above the critical size the pay-offs are reversed. Therefore, small females should be single-brooded whereas larger ones should reproduce twice within each year, and this seems to be what actually happens in the field populations that Lawlor studied.

5.7 Storage as a Special Kind of Growth

In principle, selection should favour the maximum use of energy and other resources in the production of metabolising tissue (because this should enhance growth and reduce generation time) or in the formation of gametes. Energy in excess of the requirements of somatic growth should be used in reproduction. In practice, however, some energy is stored.

Storage in the invertebrates occurs prior to the onset of very poor conditions (e.g. prior to winter or drought) in which offspring would have a poor chance of survival (Calow and Jennings, 1977) and usually occurs prior to aestivation in Mollusca and pupation in holometabolous insects. This kind of storage strategy is associated with predictable alterations in the environment or in the state of the organism. Other organisms may respond to unpredictable but probable deteriorations in food supply by carrying a store as an insurance strategy, e.g. free-living Platyhelminthes which face this kind of problem carry a store of fat whereas parasitic Platyhelminthes, which live bathed in food, do not (Calow and Jennings, 1977). Finally, energy may be stored and accumulated at one time of year ready for use at a time which is more favourable for gamete production. For example, the mussel, *Mytilus edulis*, accumulates glycogen during spring and summer when food is readily available and this is used for vitellogenesis during winter. The latter allows spawning to occur in spring and early summer.

There are differences not only in the amount of energy that is stored but in the way that it is stored. Commonly used storage materials are glycogen (polysaccharide) and lipid, though proteins are sometimes employed. Table 5.2 indicates how the use of these products is distributed throughout the invertebrates. Interestingly glycogen predominates but proteins are sometimes used and this contrasts with vertebrates where lipids constitute the major storage products.

Lipids are more efficient than glycogen in storing energy in the sense that they package more energy per gram: approximately 38 J as opposed to 17 J. However, slightly more energy is lost in the formation of lipids from fatty acids and in the transfer of this energy to ATP than in

Table 5.2: Phylogenetic survey of main energy stores.

Phylum	Store
Protozoa	glycogen and protein
Cnidaria	glycogen and protein
Platyhelminthes	
free-living	lipid
parasitic	glycogen
Nemertea	lipid
Nemathelminthes	
free-living and parasitic	glycogen
Annelida	glycogen and lipid
Mollusca	glycogen and protein
Arthropoda	lipid and protein

Source: Adapted from Calow (1981).

the synthesis and utilisation of glycogen. Furthermore, glycogen is more readily available for anaerobic metabolism. Proteins are relatively poor stores in that they package but a little more energy per unit biomass than the polysaccharides (approximately 20 to 25 $J.g^{-1}$) but incur considerably greater costs in transformation. It is surprising, therefore, that proteins are used so widely as storage products by invertebrates and that lipids have not been utilised more. These issues require more careful research and analysis.

5.8 Allometric Growth

Huxley (1932) discovered that the specific growth rate of one part of the body (y) usually stands in a constant ratio to the growth rates of other parts or to the organism as a whole (x). This is illustrated in Figure 5.10. The phenomenon is often referred to as allometric or heterogonic growth and can be represented mathematically as:

$$\frac{dy}{dt} \cdot \frac{1}{y} : \frac{dx}{dt} \cdot \frac{1}{x} = \alpha \qquad (5.5)$$

i.e. the rate of change of one organ (d/dt) stands in a constant proportion (i.e. α) to another. When $\alpha > 1$, y grows more rapidly than x (positive

Figure 5.10: Double-logarithmic plots of size of part y against size of part x at the same time – (a) abdomen breadth (y) against carapace length (x) of a shore crab; (b) total phosphorus content (y) against body fresh weight (x) of *Tenebrio;* (c) diameter of thoracic ganglion (y) against carapace breadth (x) in crabs – *Pachygrapus* (upper) *Carcinus* (lower).

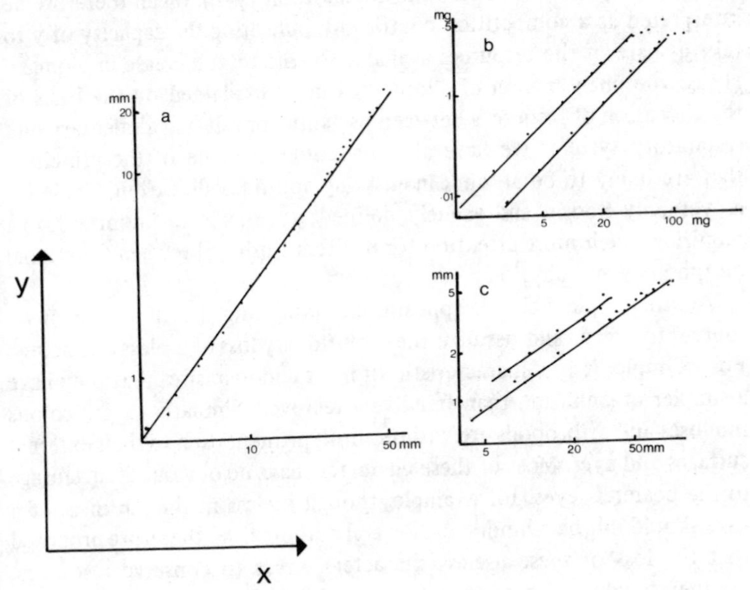

Source: All after Huxley (1932).

allometry); when $\alpha < 1$ the reverse is true (negative allometry); when $\alpha = 1$, y and x are said to be related isometrically. Integrating equation 5.5 gives:

$$y = bx^{\alpha} \tag{5.6}$$

and taking logarithms gives:

$$\log y = \log b + \alpha \log x \tag{5.7}$$

These equations are analogous to those relating food and oxygen consumption to body mass (Section 5.2).

That the specific growth rates of parts and other parts, and of parts and wholes should be related thus is not obvious. However, the functional basis of allometry can be clarified by rewriting equation 5.5 as:

$$\frac{dy}{dt} = \alpha \frac{dx}{dt} \cdot \frac{y}{x} \tag{5.8}$$

That is, part y receives from the increase of the total system a share which is proportional to its ratio to the total (y/x). α can therefore be interpreted as a competition coefficient, indicating the capacity of y to take a share of the resources available for the total increase in biomass. Hence the phenomenon of allometry can be explained on the basis of the allocation of resources between parts. In considering alimentary and respiratory systems we have already hinted at some of the principles that are likely to be at work in ensuring optimum allocation. These are as yet only poorly and vaguely defined; yet this is an important area requiring much more attention for it offers a bridge between functional morphology and physiology.

Another topic which is apposite in considering the allocation of resources to organs and tissues is the evolutionary loss of useless characters. For example, it is a characteristic of most endoparasites that they have a simpler organisation than free-living relatives. Similarly cavernicolous molluscs and arthropods are without both pigmentation to their external surfaces and eyes. Most of these characters have no obvious disadvantage to the bearer – eyes, for example, though useless in the darkness of a cave should not be a hindrance. Several authors have therefore proposed that the loss of these useless characters serves to conserve resources; i.e. they invoke the economisation principle. In other words there is no point in allocating resources to a defunct organ and, indeed, there may be a positive advantage in using the resources so saved for the construction and support of more useful (in terms of fitness) characters (Barr, 1968).

5.9 On Growth and Ageing

The idea that ageing starts when growth stops has a long and chequered history, tracing back to the work of Minot (1889). From the review given in Table 5.3 it is certainly true that ageing is most obvious in those invertebrates which suffer the strictest growth limitations (for example Nemathelminthes and Insecta) and least obvious in those which grow exponentially (e.g. Cnidaria). However, an alternative explanation for this correlation is that it depends more on the presence and absence of the turnover of tissue in adults than growth limitations *per se*. For example, cnidarians show extensive mitotic activity in all

Table 5.3: Distribution of senescence.

	Suspected lack of ageing in some spp.	Suspected presence of ageing in some spp.	Definite ageing in some spp.
Cnidaria	✓	✓	
Platyhelminthes	✓	✓	✓
Mollusca	✓	✓	
Nemathelminthes			✓
Annelida	✓	✓	✓
Rotifera			✓
Arthropoda			✓

tissues throughout life, whereas cell division and hence replacement are strictly limited in adult nematodes and insects. Interestingly, mitosis is more extensive and persists for longer in those insects with the greatest longevity, namely the Coleoptera.

The reason for proposing this negative correlation between the extent and persistence of turnover in the tissues of adults and their rate of ageing is that there is likely to be a causal link between the accumulation of damage in a tissue and its rate of turnover and between the level of damage in the tissue and its functional effectiveness or 'vitality'. Subcellular damage occurs throughout life due to thermal noise, mistakes in the synthesis of macromolecules and a variety of other processes; but this damage can also be removed by repair and replacement which depends in part on the turnover of cells. Hence the density of damage in tissues is likely to be a balance between its generation and its rate of removal. One set of theories suggests that ageing occurs because the generative processes increase in rate with age and others suggest that it occurs as a result of a reduction in the efficiency of repair processes. Both may be involved (Calow, 1978a). In any event, if and when cell turnover slows down or stops, the levels of subcellular damage will increase and ageing events are likely to manifest themselves. However, molecular turnover will also be involved as well as cell turnover and this can take place within static cells. Unfortunately the relative importance of each of these components in different organisms has yet to be established, but there is evidence that enzymes turn over more slowly in aged than in young nematodes (Rothstein, 1979).

The evolutionary question of why some systems are immortal and others are not remains unanswered, but a possible hypothesis is that both cellular and molecular turnover are expensive and that the expense is only worth bearing in systems where the chances of death by predation, accident and disease are low anyway (Kirkwood, 1980). Most animals in nature in fact die by these 'unnatural', exogenous factors. Hence, there is no call for metabolic mechanisms which confer immortality and the materials and resources that might have been expended in effecting this can be used more profitably (from the point of view of fitness) to promote survivorship and reproduction. The question of the presence and absence of ageing in animals can therefore be reduced, once again, to the relative merits of allocating limited resources between differing body needs, i.e. to the economisation principle. However, as yet, this general hypothesis does not give much guidance on particular issues; for example, why is cell constancy and ageing such an obvious feature of the Nemathelminthes and Rotifera? Why does cell turnover cease in most tissues of most adult insects? Why has investment in continuous turnover been favoured in many cnidarians? These are puzzles that continue to mock the comparative science of gerontology.

5.10 On Degrowth and Rejuvenation

Many of the lower invertebrates are capable of surviving through long periods of starvation, using their own tissues for subsistence and showing spectacular powers of shrinkage or degrowth. This has been recorded in Cnidaria, planarians, Nemertea and even some Annelida and Gastropoda (Calow, 1980). In the latter case the soft tissues shrink within the shell.

Shrunken animals often resemble juveniles and so it has been suggested that degrowth not only takes animals back in size but also in age. Dawydoff (1910) claimed that starved Nemertea could be taken back to a ball of cells resembling early embryological states, but the process did not appear to be reversible. Alternatively, starved planarians can be taken back to and beyond their size at hatching (Figure 5.11) and within certain size limits this degrowth is reversible.

Available experimental evidence suggests that rejuvenation does take place, but that the active component is regrowth and not degrowth (Calow, 1977b). There is some evidence that regrowth occurs from the proliferation of a system of stem cells, the neoblasts, though this is not altogether undisputed. However, if the neoblast interpretation is correct, regrowth will mean that old tissue is diluted with new, and that the density of damage falls (see last section) during regrowth. Similar processes may be involved in tissue replacement after fission in the

Figure 5.11: Growth, degrowth and regrowth cycles in two planarians – *Planaria torva* and *Polycelis tenuis.* Size is measured as plan area. The initial growth is sigmoid. S = starvation and is accompanied by exponential degrowth. F = re-feeding and is accompanied by regrowth.

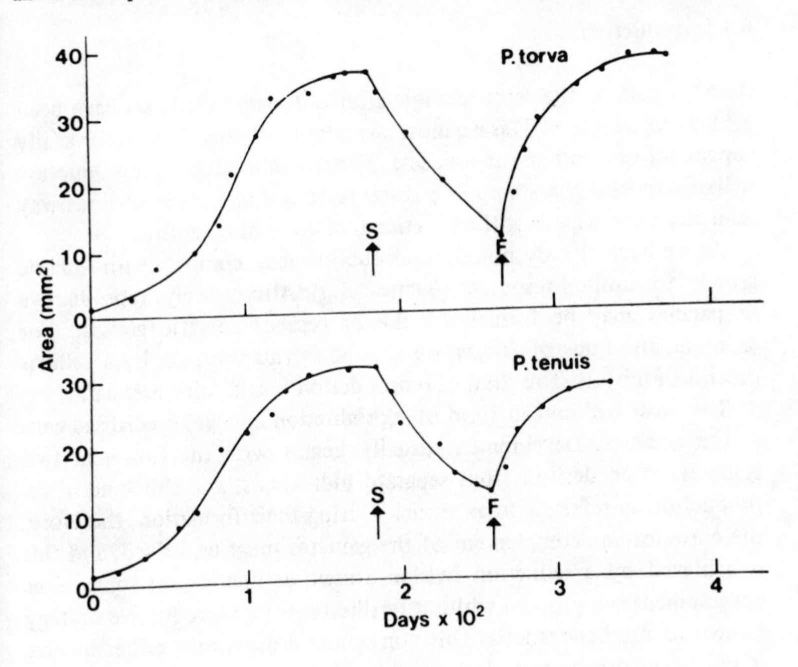

Source: With permission from Calow (1978b).

asexual reproduction of planarians and, indeed, the neoblast system may have evolved in response to the requirements of rejuvenation associated with this mode of reproduction.

6 REPRODUCTION

6.1 Introduction

Another call on the resources left after the metabolic costs have been paid is reproduction. This is a most important call since fitness is crucially dependent on reproductive success. Hence the process of reproduction is likely to bear the stamp of natural selection in a more obvious way than any other aspect of the functional biology of organisms.

As we have already noted, reproduction may compete with somatic growth for limited resources (Section 5.6). Alternatively, reproductive propagules may be formed by the process of somatic growth. For example, the buds of *Hydra* are formed in this way, i.e. by a cellular process of mitosis. This kind of reproduction is said to be asexual.

The most widespread form of reproduction involves specialised cells – the gametes. Development usually begins with the fusion of two gametes, often derived from separate individuals, and this kind of reproduction is referred to as sexual. During their formation, therefore, the chromosome complement of the gametes must be halved, and this is achieved by a reduction-division known as meiosis. In some cases development can proceed without fertilisation – a reproductive strategy known as parthenogenesis. This non-sexual process may either involve modified meiosis or no meiosis at all.

In the case of gamete formation there may be a trade-off between the size and number of gametes produced from the limited resources made available by the parent and between the extent to which a parent invests resources in reproduction and in its own soma. These are matters to consider first (Sections 6.2 to 6.6) before returning to methods of non-sexual reproduction in Section 6.7.

6.2 Sexual Gamete Production, Fertilisation and Early Development

6.2.1 Gametes

Resources devoted to gamete production can be used to form a large number of small gametes or a small number of large ones. For the male gametes (the spermatozoa) the main emphasis is on finding a female cell and fusing with it, i.e. on maximising the number of fusions. Here numbers therefore predominate over size. Spermatozoa are often not very much bigger than the haploid nuclei they carry but they usually have a highly specialised cytoplasm modified into a flagellum for locomotion and a head-apparatus, the acrosome, for penetrating the egg. This kind of sperm is found in marine Annelida and Mollusca. Platyhelminth sperm have two flagellae and a long, thin head. Insect sperm are also long and thin. The spermatozoa of some Gastropoda occur in two morphological types – ordinary, functional gametes and non-functional gametes which may serve to carry the others into the female tract or to support and nourish them. Some invertebrates produce sperm which do not have functional flagellae at all; for example nematode sperm which move under their own power by a little-understood mechanism and the sperm of many malacostrocan Crustacea which do not move at all and are often released in clumps as spermatophores. Figure 6.1 illustrates a variety of different invertebrate sperms.

Eggs are less variable in shape, usually being quite spherical, but are more variable in size than spermatozoa. As well as carrying genetic information they must also carry provisions for early development. Hence

Figure 6.1: Invertebrate spermatozoa. a = non-specialised one of an annelid or mollusc; b = malacostrocan sperm; c = planarian sperm; d = nematode sperm; e = sperm of *Drosophila;* f_1 and f_2 = respectively functional and supportive sperm of a gastropod. Not drawn to scale. Nuclear area is shown in black.

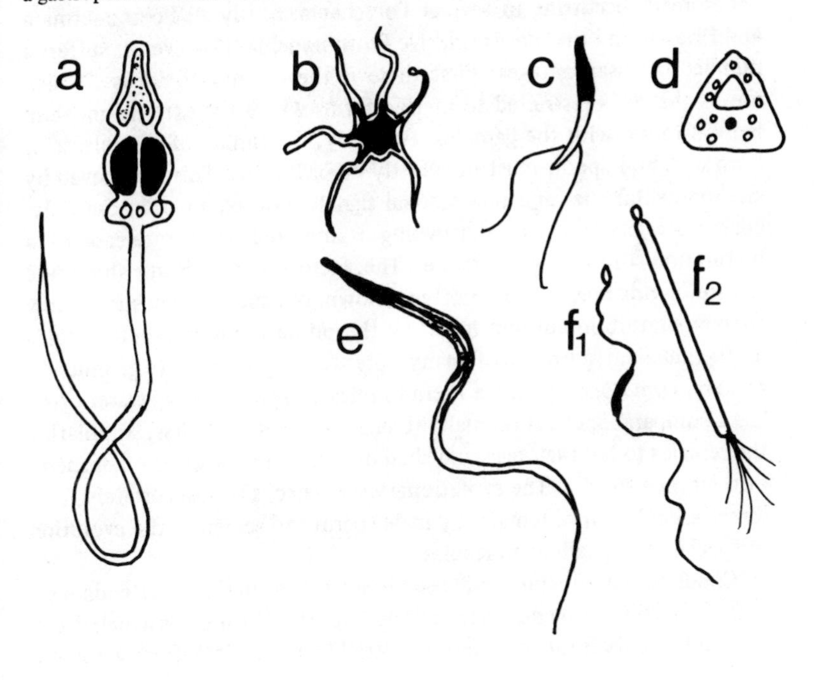

there is a real compromise here between gamete size and numbers. The nature of the compromise depends upon ecological factors and will be discussed in greater detail below. Energy resources are incorporated as fatty materials referred to generally as yolk. The eggs of gastropods contain little yolk (alecithal). Insect eggs are surrounded by yolk and are referred to as centrolecithal. In several invertebrates the yolk is supplied outside the egg (i.e. is entolecithal); for example the eggs of lymnaeid and planorbid, freshwater gastropods, float in a nutrient albumen. Most Platyhelminthes produce yolk cells in a special and extensive part of the reproductive system, the vitellarium. These vitelline or nurse cells are enclosed within the egg-shell and support development. Finally, eggs which develop more slowly than others are often eaten by the first young. These trophic eggs are common in Platyhelminthes, Annelida, Mollusca and Arachnida.

6.2.2 Fertilisation

Fusion of sperm with egg (syngamy) must occur to initiate development (but cf. Section 6.7) and the processes which bring it about are referred to as fertilisation. The product of fertilisation is the diploid zygote. External fertilisation, outside both parents, is the primitive state and is widespread, occurring in several Polychaeta, nearly all Echinodermata and Bivalvia and the most primitive Prosobranchia. However, it suffers a number of disadvantages. First there is the trauma of release, which means that it is restricted to marine forms where the external medium is iso-osmotic with the gametes. Secondly, the timing of the release of both eggs and sperm must be exactly co-ordinated. This is achieved by environmental cues and pheromonal signals. For example, in the polychaete *Arenicola marina,* spawning is initiated by the release of a maturation-stimulating hormone. The factors which bring about the release of this hormone are not yet known, but sudden exposure to low air temperature in autumn might be the prime stimulus. Seminal fluids in the inhalant currents of many filter-feeding females (e.g. mussels, oysters, *Pomatoceros*) cause them to release eggs, while egg-water often has a comparable effect on males. Usually males spawn first, stimulating the females to lay their eggs and shed directly into a suspension of sperm of their own species. The evolutionary sequence is almost certainly from direct stimulation of females by male sperm and semen to the evolution of specialised signalling molecules.

Copulation is a major tactic used in solving the problems attendant on external fertilisation and is particularly important in animals which live in the inhospitable freshwater and terrestrial habitats. Both freshwater and dry air are very effective, quick-acting spermicides. True copulation, involving the transfer of sperm from male to female without any contact with the outside, involves the development of some method of transfer. Sometimes this occurs by the hypodermic impregnation of the partner (e.g. in some Platyhelminthes which have a spined penis for this purpose) or by the male organ plugging into a specially modified female atrium. Pseudo-copulation also occurs in a number of groups. In the nemerteans *Amphiporus* and *Lineus,* males and females secrete a common mass of slime into which both sperm and eggs are released for fertilisation. Some polychaetes also use a similar strategy.

Copulation, with the transfer of naked spermatozoa, occurs in Platyhelminthes, Oligochaeta and Mollusca. Fertilisation in Crustacea involves the transfer of 'parcels' of the immobile sperm (i.e. as spermatophores) or stringy bundles from male to female by special appendages. The female uses other appendages to draw the sperm over her eggs. Male cephalopods produce enormously complex spermatophores which are syringe-like. The latter are activated by the absoption of water and pump sperm into the female without the need for male presence. Terrestrial arthropods transfer sperm in water-tight spermatophores.

6.2.3 Separate Sexes and Sex Changes

In the vertebrates, male and female gametes are most often produced by morphologically distinct male and female individuals. However, in the invertebrates it is far more common to find individuals carrying both male and female gonads and systems. This is referred to as hermaphroditism.

Hermaphroditism has a number of advantages over the separate-sexed or gonochoristic condition. For example, in sluggish animals or those which live at low population density it is likely to improve the chances of a successful mating contact. This is because all mature individuals are potentially available for fertilisation and are potentially capable of effecting a fertilisation. In agreement with this theory, simultaneous hermaphroditism occurs in a number of phyla where individuals are sluggish and/or populations occur at low densities – Porifera, Ctenophora, Platyhelminthes, oligochaete and hirudinean Annelida. Hermaphroditism is a dominant and possible primary adaptation of the largely sluggish Mollusca, yet, as would be predicted from the 'low mobility/density model', the more visually acute cephalopods are all gonochoristic. On the other hand, many opisthobranchs are rapid-movers and yet all are hermaphrodite. Similarly, some of the sluggish and immobile bivalves are gonochoristic. Perhaps in these instances population density is a more important factor than activity *per se.* For example, low population densities and poor acuity may mean that mating contacts are rare in opisthobranchs despite their high mobility. Alternatively, high population densities of even sluggish or immobile organisms will mean that mating contacts either between externally liberated gametes or individual adults are sufficient to favour gonochorism. These ideas require further investigation.

Another method of adapting to rare mating contacts is for females to store sperm. This is common throughout the Invertebrata, occurring in both bisexuals and hermaphrodites, but it is particularly common in crustacean zooplankters which often exist in sparse, thinly spread populations.

A hermaphrodite which fertilised itself would have an even greater advantage at low density, for it would then be completely independent of mating contacts. However, offspring produced by out-crossing are usually fitter than those produced by selfing. This is because inbreeding leads to increased homozygosity and so facilitates the expression of deleterious recessive genes. Hence methods have evolved to prevent self-fertilisation in invertebrate hermaphrodites. For example, various forms of *sequential* hermaphroditism have evolved where the gonad functions first as one sex and then gives way to the other; male-to-female is protandry, female-to-male is protogyny. However, under these circumstances the advantages associated with the 'low mobility/ density model' discussed above no longer apply and other advantages must be postulated. One such hypothesis is the 'size advantage model'. Suppose that the reproductive functions of one sex were better effected by a small animal and those of the other by a large one. An animal which as it grew changed sex would maximise its reproductive potential. Protandry is widespread in Mollusca in which female fecundity is related positively to body size whereas male fecundity is related to mobility and hence, possibly, small size. These differences apply particularly to sedentary species with planktonic larvae. An isolated female becomes adult when she reaches the optimum size and attracts mobile males for fertilisation. Examples occur in the Calyptraeidae (*Crepidula fornicata*), oysters, wood-boring bivalves and parasitic mesogastropods.

The disadvantages of selfing are also overcome by species with dwarf males. In these the male is much reduced and is usually permanently attached to the female. Examples are common in the Crustacea, where males may either feed independently or be parasitic on the female. The result is an effective, simultaneous hermaphrodite, in that male and female parts are permanently combined in one 'individual'. However, since male and female parts have different parents, 'selfing' can take place without detrimental effects. It could be argued that since the male and female have to come together they face the same problem as a normal gonochorist. However, associations between prospective male and female parents often occur during the larval phases and hence during a high mobility and high density part of the life-cycle.

Another disadvantage of hermaphroditism, particularly the simultaneous variety, is that each animal must build and maintain two sets of reproductive apparatus whereas gonochoristic animals support only one. Hence, each hermaphrodite has higher 'running costs' and has less energy to spend on the formation of gametes than the gonochorist (Heath, 1977). The cost of maintaining two-organ systems can be offset if one is reduced relative to the other or if there is any sharing of organ-systems. This is well-exemplified by the Mollusca in which gastropods even have a shared gonad – the ovotestis (Figure 6.2). Sequential hermaphroditism obviously avoids the disadvantage of 'higher running costs'.

Figure 6.2: Genital system of a hermaphrodite gastropod, *Physa fontinalis* – O = ovotestis; h = hermaphrodite duct; a = albumen gland; ♀ = female duct; ♂ = male duct; p = penis; v = vagina.

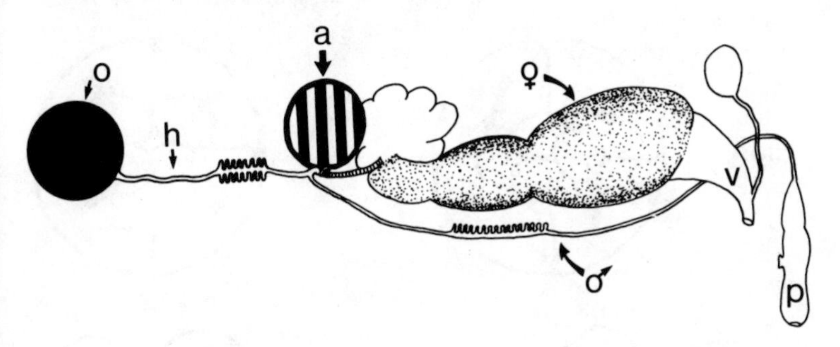

Source: After Duncan (1960). Illustration by L.J. Calow.

6.2.4 Early Development

The product of fertilisation is a single-celled zygote. During the process of development this single cell must be transformed into a multicellular organism with a complex shape, in which there is a division of labour between cells and in which cells of different types are organised into tissues and organs. Early development involves active cell division by mitosis (cleavage) and the positioning of the main germ layers (gastrulation), i.e. ectoderm and endoderm in the two-layered Cnidaria (diploblastic) and ectoderm, mesoderm and endoderm in the triploblastic higher phyla.

There is a fundamental distinction between eggs in which the ultimate developmental fate of cleavage products (the blastomeres) is determined from the beginning (determinate cleavage) and those in which fates are not fixed until a late stage in development (indeterminate cleavage). Removal of blastomeres in the former but not the latter leads to the formation of an abnormal embryo if the process gets that far. Blastomeres arising from determinate cleavage must be precisely positioned in the embryo and this is usually associated with a very ordered method of cleavage known as spiral cleavage. Precise positioning is not so essential in the indeterminate process and here a less ordered process of radial cleavage takes place. Determinate and spiral cleavage is typical of Platyhelminthes, Nemertea and Annelida. Arthropods also fit into this category but here the cleavage processes are obscured by the large mass of yolk in the egg. Cnidaria and Echinodermata have a less-determined egg and show simple radial cleavage. Examples of these two processes are illustrated in Figure 6.3.

Figure 6.3: Spiral and radial cleavage at the four-cell stage.

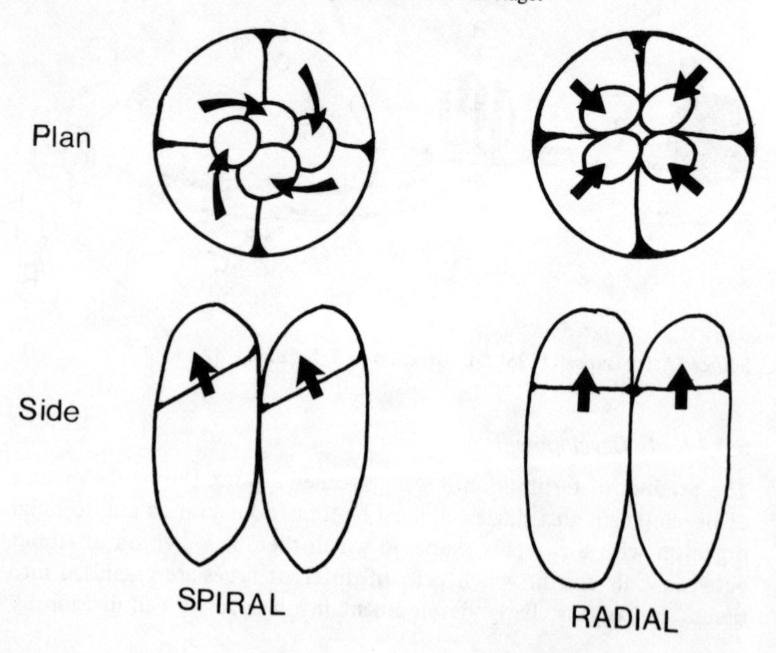

Nemathelminthes show a very peculiar method of cleavage with cells moving over each other, and division occurring at odd, but precisely determined, angles to each other. As already noted each organ of a nematode ends up with a very exact number of cells and any cell removed or killed during development leaves a precise gap.

Usually the end-product of cleavage is a hollow ball of cells, the blastula. Positioning of the major germ layers then involves cell movements – i.e. cells migrate into the blastocoel (inner cavity of the blastula) to form a gastrula. Thereafter organs and tissues form from the major germ layers of ectoderm, endoderm and, in triploblastic animals, mesoderm, as indicated in the summary Table 6.1. This involves differentiation and movement of cells to form tissue and organ patterns. Shaping, or morphogenesis, is also achieved by the differential growth processes discussed in Section 5.8.

Table 6.1: Fate of major germ layers.

Germ layer	Derived tissue
Ectoderm	Epidermis Nervous tissue Linings to buccal cavity and rectum Some excretory systems
Endoderm	Gut Gut glands Part of the gonads of diploblastic animals
Mesoderm	Muscles Excretory systems Part of the gonads of triploblastic animals Some excretory systems

6.3 Marine Life-cycles and the Trade-off between Egg Size and Numbers

Marine invertebrates exhibit a wide range of life-cycles. However, there are two principal types. In some species benthic adults produce planktonic larvae which undergo metamorphosis in conjunction with settling. Most of these planktonic larvae feed in the plankton (i.e. are planktotrophic), but others do not. The others carry a yolk store with them and are large. They are said to be lecithotrophic. The other principal group produces offspring which develop directly into adults without metamorphosis.

Figure 6.4 shows some examples of planktotrophic and lecithotrophic larvae. The former are smaller but more complex than the latter. Hence more planktotrophic larvae can be produced per unit resource available for reproduction than in lecithotrophic species. The price that must be paid for this advantage is dependence on an external food supply and possibly slower developmental rates, which means that planktotrophic larvae remain in the plankton for longer than lecithotrophic larvae and are thus subjected to the dangers there (e.g. predators) for longer. Direct development, as occurs in many littoral invertebrates, avoids the latter but may be even more expensive because energy has to be set aside for either egg-case formation or brooding. Furthermore, direct, benthic development loses any advantages that might arise from dispersal. From these generalisations it is not difficult to formulate a number of predictions on the distribution of life-cycle types in marine invertebrates (and these have been established rigorously by Vance, 1973a, b and Christiansen and Fenchel, 1979; but cf. Underwood, 1974). The more obvious ones are as follows:

(1) When food is abundant and predation in the plankton is low, planktotrophy should be favoured.
(2) When food is rare and/or predation is more intense, lecithotrophy should be favoured.
(3) When planktonic mortality is greater than benthic mortality, direct development should be favoured.

Some of these predictions are borne out by observations on the geographical distribution of developmental types. For example, in the arctic and antarctic seas non-pelagic (direct benthic) development is dominant and Thorson (1950) has correlated this with low primary productivity and low temperatures. It is true that the antarctic waters are noted for their high levels of productivity but this enormous production occurs at the very surface of the open ocean, whereas most invertebrates are restricted to the shallow-water shelves of the coast.

When, however, a larva is able to live under the severe conditions of these polar waters, the planktotrophic state seems to afford a good opportunity in competition with those species which have non-pelagic development. More than 95 per cent of the species in the arctic and antarctic have a non-pelagic development but the 5 per cent reproducing with planktotrophic larvae comprise several of the species which quantitatively are commonest in the area, for example *Saxicava arctica* (Annelida), *Mya truncata* (Mollusca) and *Strongylocentrotus*

Figure 6.4: Examples of planktotrophic and lecithotrophic larvae. The names given to the various kinds of larvae are: a = Müller's; b = pilidium; c = Iwatas; d = trocophore; e = trocophore; f = veliger; g = trocophore; h = pluteus; i = diplura.

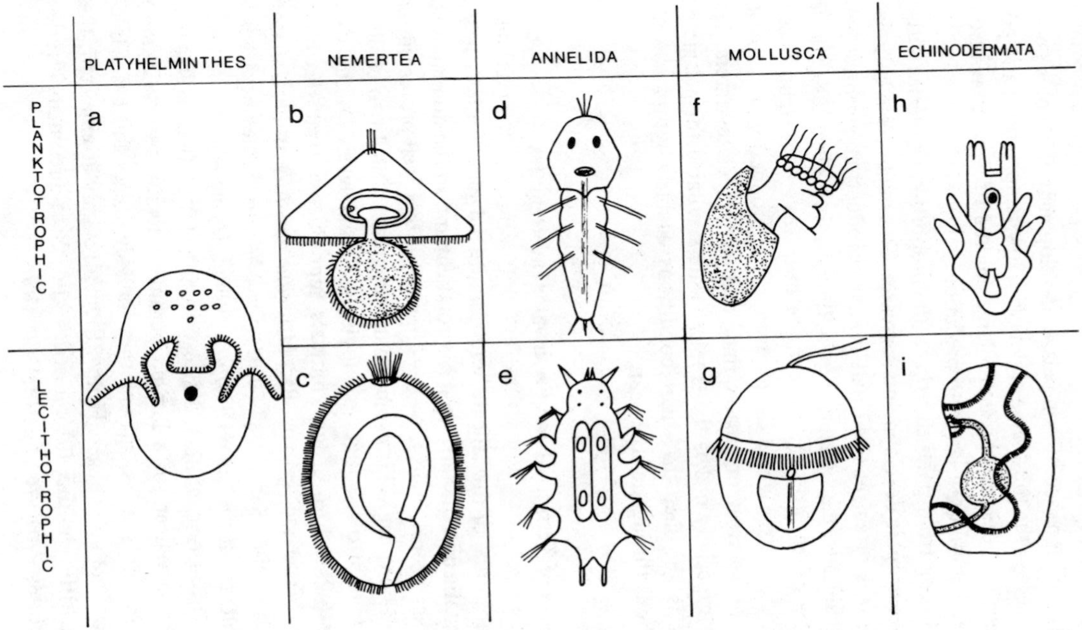

	PLATYHELMINTHES	NEMERTEA	ANNELIDA	MOLLUSCA	ECHINODERMATA
PLANKTOTROPHIC	a	b	d	f	h
LECITHOTROPHIC		c	e	g	i

Source: Illustration by L.J. Calow.

droebachiensis (Echinodermata).

Curiously, lecithotrophic larvae are absent or nearly so from polar waters where their independence from food and their ability to spread would seem to give them an advantage. The reasons for this are not very clear. Thorson (1950) thinks that the very harsh conditions in these environments preclude compromises and favour only the extreme strategies of large eggs and non-pelagic development or small eggs and pelagic development.

The dominance of non-pelagic development also holds good for the deep seas – the third large region with low temperatures and poor food. Ascent and descent through a large column of water may also increase the vulnerability of larvae to predation.

Moving from poles to tropics, the proportion of species with pelagic, planktotrophic developments increases. This might seem curious at first, since it is often reported that primary production is greater in temperate than in tropical waters. Again, though, such a conclusion is based on studies on surface waters over the deep sea – not upon observations from the shelves where the phytoplankton production of tropical waters may be as great or greater than that of temperate waters. However, what especially makes the tropical waters suitable for pelagic development is the fact that light conditions enable phytoplankton photosynthesis throughout the year.

6.4 Eggs of Terrestrial and Freshwater Invertebrates

Larval forms, of the delicate, undifferentiated, planktonic variety cannot be sustained in the terrestrial environment and so the dominant strategy here has been to telescope development into a well-protected, cleidoic (closed-box) egg. This evolutionary progression ought to be discernible in extant littoral species occupying different levels on the shore and Table 6.2 compares and contrasts reproductive strategies for such a series of related gastropods. *Littorina saxatilis* is best adapted for terrestrial life. Surprisingly *Littorina obtusata* comes next best while the salt-marsh species, *Melampus,* though laying egg masses which give some protection against desiccation and destruction, still has a free-swimming veliger for a limited period of its life-history. *Acmaea,* the most marine species, spawns tiny individual eggs and has a free-living trochophore and veliger. Hence, though roughly as expected, there are irregularities in this littoral series and these presumably reflect the opportunistic nature of natural selection.

Table 6.2: Vertical zonation in intertidal snails and their reproductive patterns.

Species	% time submerged	Eggs	Young
Melampus bidentatus	5-<2	egg masses	veliger
Littorina neritoides	5-<2	egg capsule	veliger
Littorina saxatilis	30-5	viviparous	miniature adult
Littorina littorea	70-25	egg capsule	veliger
Littorina obtusata	85-55	egg masses	miniature adult
Lacuna vincta	100-85	egg masses	veliger
Acmaea testudinalis	100-85	individual eggs	trocophore

In fully terrestrial forms, however, encapsulated, cleidoic eggs are the rule – i.e. capsules and cocoons are common in terrestrial 'worms', molluscs and insects. Associated with 'closed development' eggs must be provisioned and so the general emphasis in these animals has been towards large eggs rather than large numbers of eggs. A few species retain eggs either in a specialised part of the female reproductive tract (e.g. some insects) or in external brood pouches (e.g. isopods), but there is rarely any direct connection between parent and offspring of the kind found in viviparous vertebrates and so even here large sized eggs are favoured.

Some invertebrates have arrived in freshwater habitats via terrestrial ancestors whereas others have entered freshwaters via estuarine ancestors. Freshwaters themselves, however, are unsuitable for larval development due to strong currents, osmotically dilute conditions and poor mineral supplies, and so the same trends in reproduction are observed in all freshwater invertebrates (irrespective of origin) as in terrestrial species.

These trends are well-exemplified by freshwater gastropods. Pulmonate gastropods have a terrestrial origin and prosobranchs have invaded directly via estuaries. Egg-capsule production is dominant in the pulmonates and here the size of individual eggs tend to 1 mm in diameter and the number of offspring per parent per breeding season is between 100 and 1,000. Several to many eggs are packaged in each capsule and the cost of producing the capsule wall itself may not be insignificant; for example, approximately one-third of the ash-free dry mass of the egg capsule of the river limpet is capsule membrane and has an energy-equivalent of about 19 J/g as opposed to the value of 25 J/g for the capsule and its contents. Some species of freshwater prosobranch produce capsules but ovo-viviparity (eggs retained in a brood pouch with yolk and no further nutrient supply from the parent) is more common. Here the constraints of brooding (i.e. the size of the pouch) mean that even fewer eggs are produced per parent than by the pulmonates. In contrast most marine prosobranchs produce smaller eggs (10-10^2 μm diameter) but more of them (as many as 10^6 per parent).

The production of capsules or cocoons which are usually fixed firmly to the substrate is typical of freshwater 'worms', whereas brooding dominates the reproduction of Crustacea. A few species of planktonic freshwater crustaceans give rise to well-developed nauplius larvae (Figure 6.5a) which have a short life in the plankton; for example, leptodorids and copepods have this stage. One species of bivalve mollusc, *Dreissenia polymorpha,* also produces free-living larvae, though this is a recent colonist of freshwater, and other species, notably the swan mussel *Anodonta* and the freshwater mussel *Unio,* produce glochidia larvae (Figure 6.5b) which become parasites of fishes before being released as miniature adults. Most other freshwater bivalves brood their young and release them as miniature adults.

Figure 6.5: (a) Nauplius of a freshwater crustacean; (b) glochidium of a freshwater bivalve.

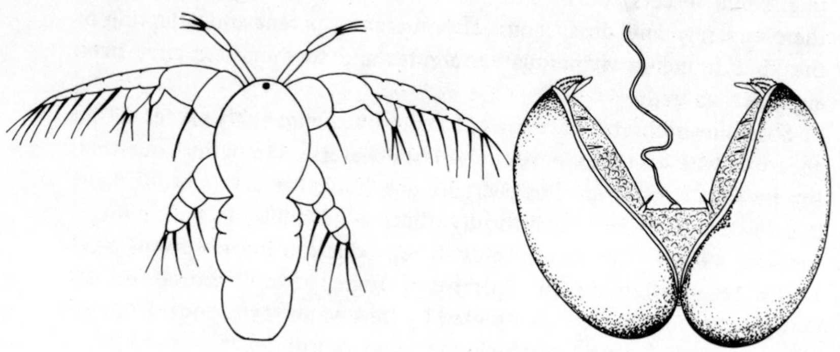

Source: Illustration by L.J. Calow.

6.5 The Complex Insect Life-cycle

Primitive wingless insects (Apterygota) hatch as miniature adults but without gonads and genitalia. Examples are the bristle-tails (Thysanura) and spring-tails (Collembola). In winged species (Pterygota), the wings of some develop externally (Exopterygota) and at each ecdysis the juveniles (known as nymphs or naiads) become more and more like the adult. This is known as hemimetabolous development and familiar examples are the mayflies (Ephemeroptera), dragonflies (Odonata), locusts (Orthoptera) and bugs (Hemiptera). In more advanced orders, wing buds are concealed beneath the cuticle (Endopterygota) and the juveniles (larvae) differ radically from the adults (imagos). There is an inactive and non-feeding stage interposed between the larval and adult stages (pupa). This is referred to as holometabolous development and examples are butterflies and moths (Lepidoptera), flies (Diptera), beetles (Coleoptera) and bees and wasps (Hymenoptera).

It is no accident that the evolution of the complex life-cycle and the wing are correlated. The evolution of the wing enabled the adult to become specialised for dispersal (cf. the marine invertebrates where it is the larvae which become specialised for dispersal) whereas the juveniles become specialised for rapid growth and development. These two processes are almost incompatible. First, flight is a very active process (see Section 3.6) and would compete with the synthetic processes for resources. Secondly, increasing the efficiency of the wing has involved simultaneous strengthening and lightening and the latter has been achieved by degeneration of living epidermal tissues which often contain much water (Maiorana, 1979). Hence, wings are incapable of being regenerated and, except in the Ephemeroptera, no growth or ecdysis occurs in any insect order once functional wings have been formed. However, ecdysis seems to have occurred in some primitive, fossil, insect adults which probably had poor powers of flight (Maiorana, 1979).

Once these distinctions had occurred between larval and adult forms, the two occupied different ecological niches were subject to different selection pressures and therefore diverged further in morphology. Larval forms, for example, range from the little-specialised oligopod larvae with three pairs of legs (as found in scarabid beetles), through the many-legged (polypod) caterpillars of Lepidoptera, to the legless (apodous) grubs of the Diptera and Hymenoptera.

6.6 The Cost of Reproduction for Parent Survival (Iteroparity versus Semelparity)

It is possible to distinguish between two major classes of life-cycle on the basis of breeding frequency and the life-spans of parents. In some species, parents breed once and die whereas in others they breed repeatedly. These are referred to respectively as semelparous and iteroparous patterns. The semelparous pattern often, though not always, occurs over one year (i.e. is annual) whereas the iteroparous pattern often, though again not always, repeats itself annually, over several years (i.e. is perennial). Sometimes, repeated breeding can occur at short intervals and these may fuse to form a long continuous breeding programme. The various patterns are schematised in Figure 6.6.

Figure 6.6: Life-cycle patterns. (a) semelparity; (b) iteroparity; (c) and (d) are the same as (b) but with varying degrees of fusion of breeding periods. ● = breeding; ★ = newborn; - - - - = release of offspring.

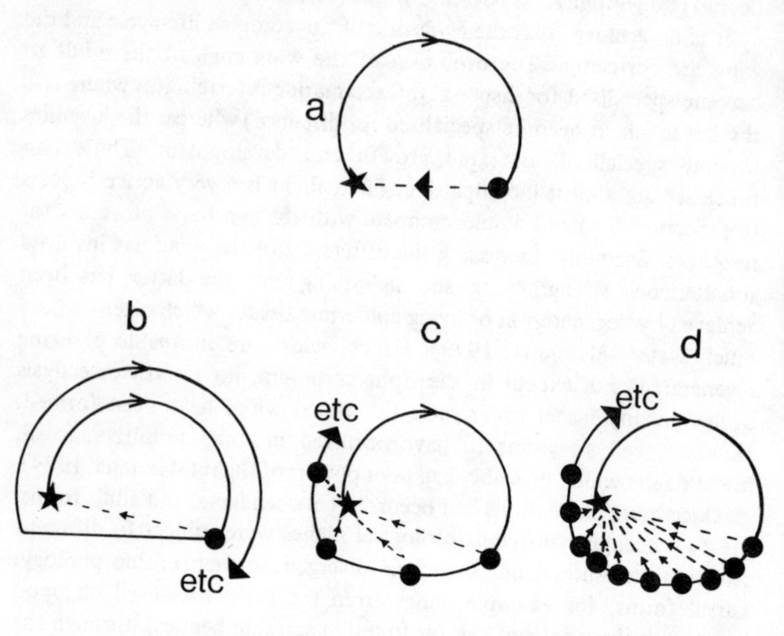

Semelparity and iteroparity are distributed over a wide range of invertebrates and even closely related species may show opposite strategies. In freshwater flatworms, dendrocoelid triclads tend to be annual and semelparous and the planariid triclads perennial and iteroparous. However, in Britain, *Planaria torva* (as the name implies, a planariid), is annual and semelparous. (These facts are reviewed in Calow, Davidson and Woollhead, 1981.) In the annelids, semelparity is characteristic of nereid polychaetes and some leeches. However, most polychaetes are iteroparous as are some leeches. Semi-continuous iteroparity is characteristic of oligochaetes and small polychaetes. (For a review see Olive and Clark, 1978.) In the Mollusca some species are semelparous and short-lived e.g. temperate, freshwater pulmonates and octopods, and others are iteroparous and long-lived, e.g. many bivalves (see Comfort, 1957). In the Arthropoda, many insects are annual and best described as semelparous or as breeding continuously when conditions allow, but some coleopterans are iteroparous and live over several years. Brooding crustaceans best fit into the pattern of continuous iteroparity. Some spiders die after producing one brood in a season, others produce several broods and yet others, the large tropical species, breed over several seasons. Finally semelparity and iteroparity occur patchily throughout the Echinodermata.

There appear, therefore, to be no general phylogenetic constraints operating on semelparity and iteroparity. Instead, which strategy has evolved has depended more on the particular selection pressures associated with the habitat and niche of the organisms. Under non-limiting conditions the gains to the rate of increase of a strategy in which the parent survives breeding as compared with one in which the parent is sacrificed is very little (just one extra head; Cole, 1954) – so why be iteroparous? The answer to this question, often referred to as Cole's result, is that conditions are not always non-limiting and where competition and predation are important, parents often fare better than their offspring. In other words, survival of the parent provides insurance against death of offspring – so why not always be iteroparous? The answer to this second puzzle is that adult survival can only be ensured if reproductive levels are not allowed to become too great. There is, in other words, a trade-off between the production of progeny and parental survivorship. There is also a trade-off between reproduction and the subsequent growth of parents and this has already been discussed (Section 5.6).

Evidence that reproduction takes a toll on the well-being of parents is extensive and has been reviewed by Calow (1979). Negative correlations occur, for example, between reproductive output and the length of life of some species of insects and rotifers. Furthermore it has been possible to extend the lives of semelparous invertebrates by artificially preventing them from reproducing. Virgin insects often live longer than mated, and mutant, ovariless *Drosophila* live longer than non-mutant counterparts. Virgin slugs of the genus *Agriolimax* also live longer than mated ones and selfing, which results in lowered fecundity, is less severe on adult survivorship than out-crossing.

Potential causal links between adult survival and reproductive output have also been discovered. As is shown in the comparison between the energy budgets of semelparous and iteroparous triclads at various levels of ration in Figures 6.7 and 6.8, the reproductive processes 'steal' more and more resources from the metabolism of the semelparous but not iteroparous animal as ration is reduced. As a result the semelparous parent shrinks (degrows) to support gamete production and associated with this process is an increased level of mortality.

During starvation, mated, female corixid bugs will use material from their own flight muscles to sustain egg output. These flight muscles cannot be regenerated and so the parent is irreversibly weakened by its reproductive activities. Under these conditions virgins live longer than mated females (Calow, 1979). Similarly, radiotracer work on *Octopus* has shown that during the reproductive period, amino acids are drawn from the muscles of the soma to support synthesis in the gonads (O'Dor and Wells, 1978). Finally, the diversion of energy and resources from the somatic tissue to the gonads may accelerate the ageing processes in

Figure 6.7: Energy partitioning in reproductive adults of an iteroparous triclad *Polycelis tenuis* on different rations. FSI = food supply index (and is an index of ration, which reduces from left to right). I = ingested energy. However, the absorption efficiency in this group is greater than 90 per cent. Hence I ≙ A = energy available for metabolism, and this accounts for all metabolic requirements; i.e. there is little degrowth (Dg). Note that after the ration level is reduced by 50 per cent the investment in reproduction reduces markedly.

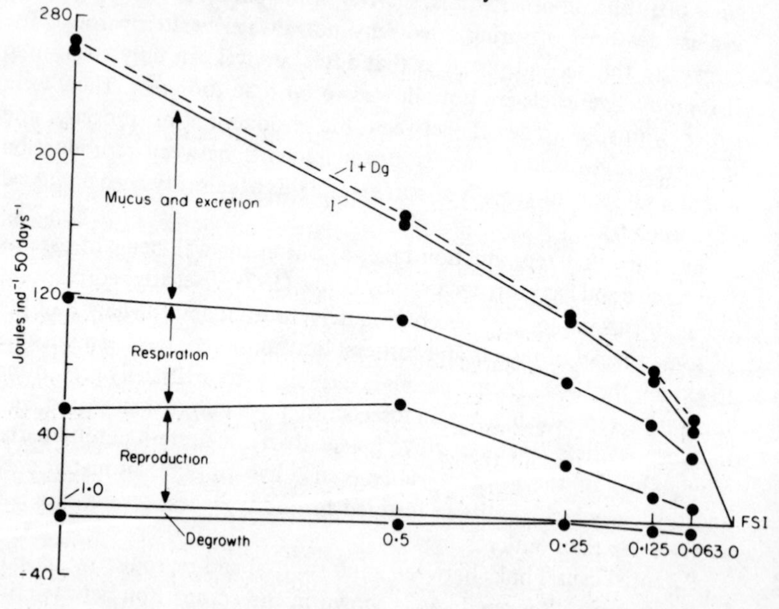

Source: With permission from Woollhead and Calow (1979).

Figure 6.8: As in Figure 6.7, but for a semelparous species, *Dendrocoelum lacteum*. Note that at no ration level was I sufficient to meet the metabolic demands. I was therefore supplemented with energy from degrowth. Reproductive investment is independent of ration and remains high throughout.

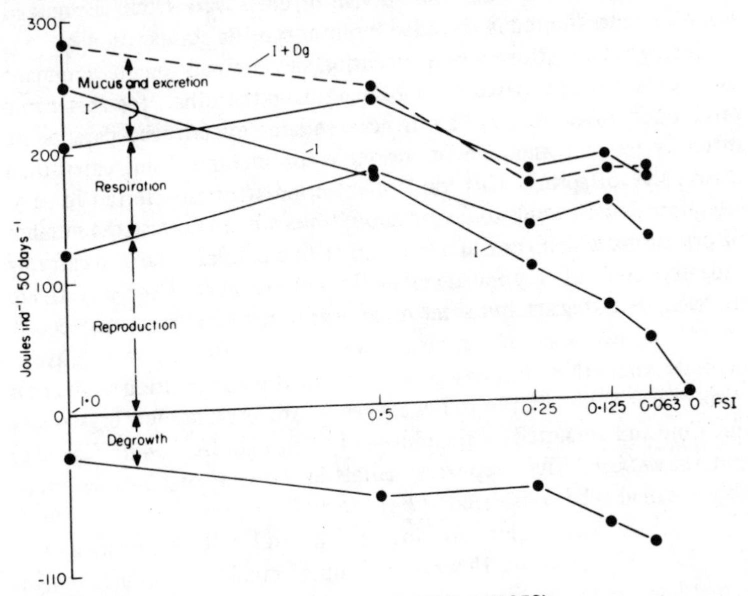

Source: With permission from Woollhead and Calow (1979).

the soma by reducing rates of turnover and repair in these tissues (Section 5.9).

Semelparity is favoured in conditions where there is a premium on population expansion – e.g. when populations are continuously reduced to low levels by inclement physical conditions or where new areas of habitat become available. This kind of natural selection is often referred to as ***r***-selection, because it favours traits which maximise the population coefficient of exponential growth, ***r***. In other words it favours the production of progeny even at the expense of the parent. *Stylophora pistillata* is an important scleractinian coral in the Gulf of Eilat in the Red Sea and has all the characters that might be expected of an *r*-strategist: small body size, rapid development, high fecundity and short life-span (Oya, 1976). Most of the other corals in this region have the opposite characters. Habitats on the reef flats near the surface are subject to the vagaries of wave action and may be exposed and dry out for long and, in this region, unpredictable periods of time. Hence, corals which occur in such habitats suffer density-independent mortality and may have to re-colonise areas denuded by emersion. Of all the corals in the Red Sea, *S. pistillata* is the one best capable of surviving in these extremes. The vast majority of the others are limited to deeper waters. Alternatively, *S. pistillata* can survive in deep waters but as soon as space becomes limited is excluded by inter-specific competition.

Ecological situations where predation and competition are dominant usually favour the persistence of parents rather than the 'big bang' production of offspring. This is because large experienced parents are often better at competing for resources or avoiding being eaten than their naive offspring. This kind of selection is often referred to as *k*-selection since it tends to favour those traits which increase the number of organisms which can survive together in a habitat – i.e. the carrying capacity or *k* of a population in the environment. Most insects are probably *r*-strategists, but some do appear to have *k* characters. According to Southwood (1976) this is true of tropical butterflies which live in habitats where there is heavy predation and also competition from other herbivores. Adults of the following genera are large, lay few eggs at any one time and are noted for their longevity: *Morpho, Heliconius, Charaxes* and *Hamadryas.* The deep-water corals in the Gulf of Eilat, described above, can also be considered as *k*-strategists.

Obviously *r* characters are not very helpful in *k* circumstances of selection and *vice versa.* However an unpredictable environment might sometimes favour parents if they stand a better chance of surviving the inclement interludes. This strategy is often referred to as *bet hedging.* The pond skater, *Gerris thoracicus,* shows delayed reproduction, small egg batches, iteroparity and a long life-expectancy of the imago (Vepsälänen, 1978). At the same time it lives in harsh, transient habitats and larval survival is low. Iteroparity here, therefore, insures against the high and unpredictable death of offspring. A similar example is found in the British millipedes. Here there are semelparous and iteroparous species and on average the semelparous species produce more offspring per parent per breeding season than the iteroparous. Blower (1969) found that the semelparous species he studied were evenly distributed amongst the soil and litter of deciduous woodland, whereas the iteroparous species tended to have more specialised requirements and were more aggregated. His semelparous species fed on leaves which were evenly distributed, but at least one of the iteroparous species fed and laid eggs on more patchily-distributed logs. Dispersal of young between the resource patches was a hazardous process and so Blower concluded that here iteroparity was a bet-hedging strategy selected to compensate for the unpredictability of dispersal. Iteroparity, with continuous or semi-continuous breeding, has presumably evolved for similar reasons in endoparasites but here, because the parent is surrounded by super-abundant food resources in the form of the tissues of its host, it can produce vast quantities of progeny without adverse influence on itself. In this situation, therefore, there need not be a negative relationship between reproductive output and the survival of the parent. Tapeworms are a classical example. The rodent tapeworm *Hymenolepis diminuta* may produce more than 1,000 eggs per day and devote as much as 35 to 40 per cent of its food supply to reproduction but it lives as long as its host – and even longer if it is transferred from one host to another before the host dies of old age!

6.7 Reproduction Without Sex

Reproduction occurs without sex in the invertebrates in one of two ways: first, new individuals may be produced from unfertilised eggs (parthenogenesis) or, secondly, new individuals may be formed from outgrowths of the parent (asexual reproduction).

6.7.1 Parthenogenesis

There are two types:

(1) Arrhenotoky – where the males are produced from unfertilised eggs. This occurs in social Hymenoptera in which males are haploid and females, which derive from normally fertilised eggs, are diploid.

(2) Thelytoky – where females are produced from unfertilised eggs. There are two main types:

(a) *Apomixis* – in which meiosis is suppressed and there is a single mitotic maturation. Offspring are therefore approximately identical with each other and their parents – mutation being the only source of variation. This is widespread in invertebrates, having been discovered in Nemathelminthes, Oligochaeta, Mollusca and Arthropoda.
(b) *Automixis* – in which meiosis is normal, producing four haploid nuclei from each gametocyte, but in which diploidy is restored either by the fusion of two pronuclei or of two early cleavage nuclei. Here variation is introduced by segregation and crossing-over, so that offspring may differ from their parents and from each other. However, the fusion of these nuclei will increase homozygosity

and since this increases the chances that usually recessive, deleterious genes are expressed, automixis is not widespread. It has, nevertheless, been recorded in several species of arthropods: a few species of coccid bug, one species of white fly and one species of mite. Another possible type of thelytoky involves the pre-meiotic duplication of chromosomes rather than the post-meiotic fusion of nuclei. This has been discovered in some parthenogenetic earthworms and in the Australian grasshopper, *Moraba virgo*.

Thelytoky may be obligate, facultative, cyclical or a rare aberration and some examples will now be given of each of them:

(a) *Occasional:* virgin and usually sexual stick insects, locusts and grasshoppers sometimes lay eggs which hatch spontaneously. In *Drosophila* it has been possible to increase the frequency of the hatching of unfertilised eggs by artificial selection. This kind of parthenogenesis seems to be mainly by automixis.
(b) *Obligate/facultative:* the former is usually associated with an odd and the latter with an even chromosome number. Odd chromosome numbers present problems for meiosis and here apomixis is usual. The British snail *Potamopyrgus jenkinsi* is thought to be triploid or tetraploid and is almost certainly an obligate parthenogen – only a few males have ever been described. *P. antipodarium,* a closely-related species from New Zealand, is diploid and is a facultative parthenogen, producing males when mated.
(c) *Cyclical:* among the rotifers one order, the Mongononta, are cyclical parthenogens. They inhabit freshwaters. All-female populations reproduce by parthenogenesis apomictically and can persist indefinitely. However, at high densities and low food supplies, sexual females are sometimes produced parthenogenetically. If these females are not fertilised they lay unfertilised eggs which develop into haploid males. If they are mated they lay winter eggs with resistant coats which may lie dormant for many months until better conditions return. When this happens they hatch into a new strain of parthenogenetic females. Similar cyclical parthenogenesis occurs in *Daphnia* and gall wasps (Cynipidae). The aphids show an even more elaborate mechanism, having not only ecologically-distinct sexual and parthenogenetic phases but also several ecologically and morphologically distinct parthenogenetic stages (Figure 6.9).

It can be argued that thelytokous parthenogenesis is a far more

Figure 6.9: The complex life-cycle of a bird cherry-oat aphid. This switches habitats during its life but the main point is that stages A to F are parthenogenetic morphs whereas G and H are sexual.

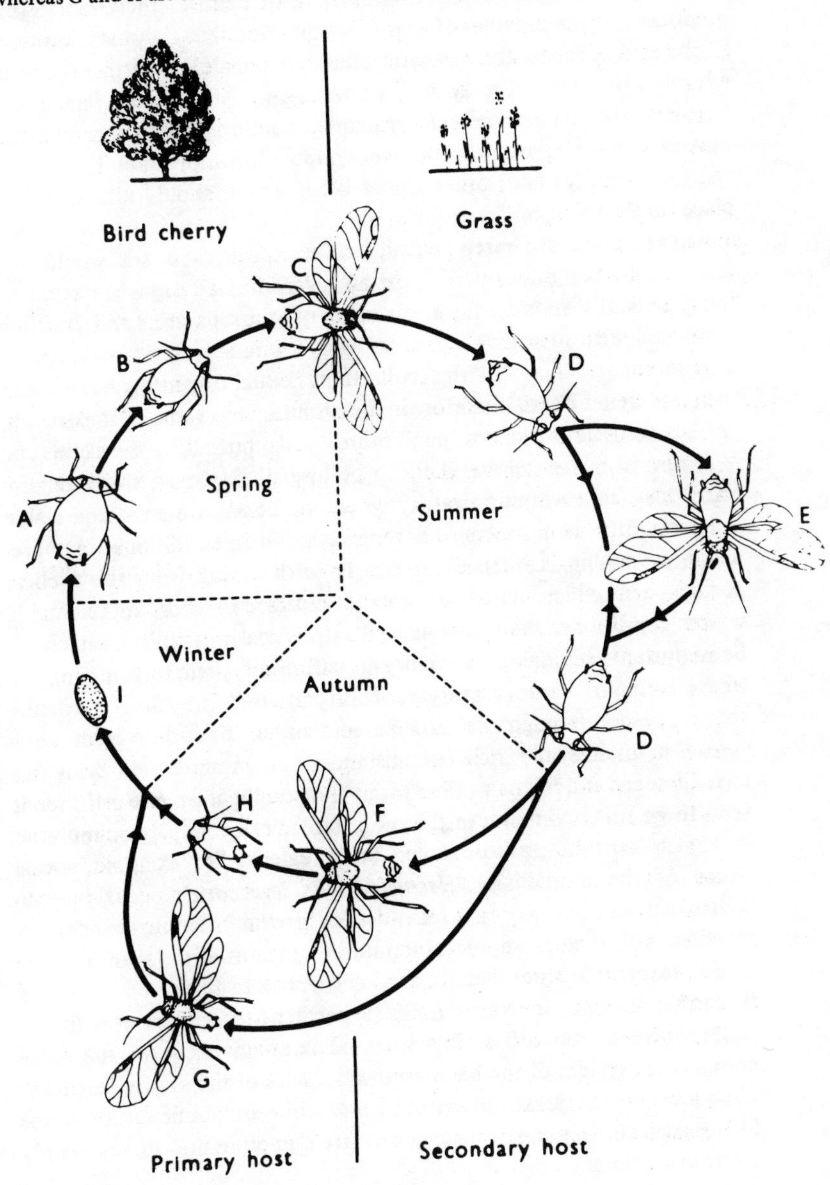

Source: With permission from Dixon (1973).

efficient method of reproduction than any sexual equivalent (Maynard Smith, 1978). Consider, for example, two females, each with the same physiology, but one with a mutation which causes thelytoky. Both produce the same number of eggs from the food that they eat. However, all the eggs of the parthenogenetic female are female and parthenogenetic whereas, on average, only half of the eggs of the sexual female are female – the rest are males. In principle, therefore, the parthenogenetic mutant should spread through the population more rapidly (in fact twice as rapidly) and, other things being equal, should ultimately replace the sexual form.

Other things are rarely equal, however, otherwise sex would not exist at all. Sex brings with it certain overriding advantages related to the genesis of variety by the mechanisms of gene separation and shuffling associated with meiosis – i.e. crossing-over and segregation – and with gene mixing associated with fertilisation. Sexual organisms 'play' more than one 'genetic card' therefore in the evolutionary 'game'. Alternatively parthenogenetic organisms 'play' more 'cards' but all of the same type – if one is a loser all are likely to be losers. 'Playing' a single genetic 'card' may be a winning strategy when conditions are good and stable but variability would seem to have the edge when conditions are unpredictably variable. Evidence for this is forthcoming from the cyclical parthenogens which switch to sex in anticipation of poor, unpredictable winter conditions. Biological as well as physical variability is likely to be important. Biological variability in pathogens, predators or competitors will require compensatory variability in hosts, prey and co-existing species. Hence it might be anticipated that sex should be most widespread in biologically rich circumstances and in agreement with this idea Glesener and Tilman (1978) find that parthenogenetic arthropods tend to be most common in the more biologically simple communities in higher latitudes and/or cooler, drier regions. For example, sexual forms of the ixodid tick *Haemaphysalis longicornis* occur only in southern Japan and Korea, south of their parthenogenetic conspecifics in ecologically more simple communities in northern Japan and the USSR. Many millipedes that are parthenogenetic in Finland are bisexual in central Europe. In North America parthenogenetic crickets live in simple cave communities, but bisexual relatives occur in the more complex ecosystem of the Mediterranean. Males of the sawfly *Leucospis gigas* are extremely rare in central France but are abundant along the Mediterranean. Examples such as these are numerous and there are only a few exceptions.

6.7.2 Asexual Reproduction

Asexual reproduction, here, refers to reproduction without gametes. It usually occurs by means of the separation of a fragment of tissue from the parent. This fragment is formed by the normal process of somatic growth and therefore by mitosis. After separation it continues to develop by mitosis. Hence, apart from the occasional mutation, the offspring formed by this process are genetically identical with their parents.

The fragment may be a non-specialised, non-individuated part of the parent as occurs in the laceration of strips of tissue from the pedal areas of Anthozoa and the simple fission of ctenophores, triclad Platyhelminthes, nemerteans, oligochaetes and some echinoderms (asteroids, ophiuroids and holothurians, but not crinoids and echinoids). After separation, these develop into miniature adults by tissue reorganisation. In other cases, accelerated development causes the fragments to take on the form of a new individual *prior to* separation. This occurs in the budding of *Hydra,* the fission of rhabdocoel Platyhelminthes and the fission of some oligochaetes. Here fission products may themselves fission prior to separation leading to the formation of chains of zooids and being the basis of colony formation in the Cnidaria. The process of separation *before* development is referred to as architomy whereas the process which involves development before separation is referred to as paratomy. The two processes are illustrated in Figure 6.10.

Two unusual forms of asexual reproduction are found in the *Porifera* and Digenea (parasitic Platyhelminthes). In the former, gemmules or reduction bodies are formed from amoebocytes which gather from all parts of the sponge to form a ball of cells. All amoebocytes are thought to be derived by mitosis. The outer amoebocytes in the ball secrete a resistant membrane. Many gemmules are formed by a single parent and are released after the parent disintegrates at the end of a season. They are resistant to harsh conditions but will ‘germinate’ when conditions ameliorate. At this time amoebocytes flow out of a micropyle, differentiate and organise themselves into a new sponge. Subsequent growth is by mitosis.

In Digenea the life-cycle is complex, involving a number of larval forms (Figure 6.11). Two hosts are involved; for example in *Fasciola hepatica,* the liver fluke, the primary host is the sheep which harbours the adult and sexual phase whereas the secondary host is a snail and harbours larvae. Hence there must be a transfer between hosts and this is likely to be a hazardous process involving heavy losses. Possibly to compensate, the larvae multiply themselves by asexual reproduction.

Figure 6.10: Fission by (a) architomy and (b) paratomy.

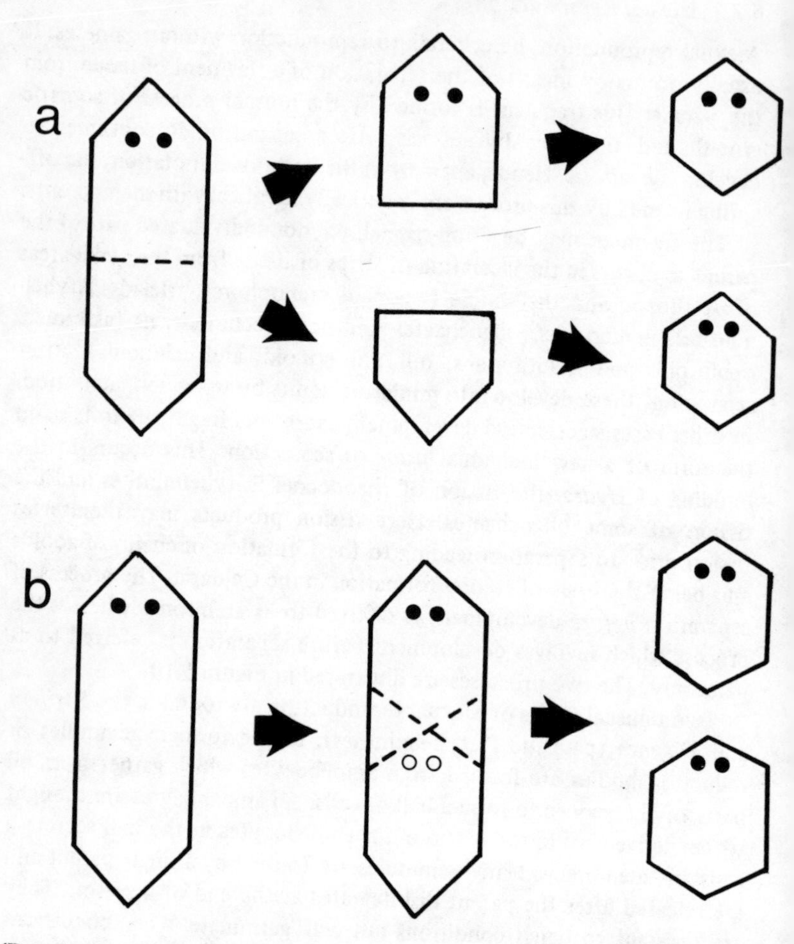

'Daughter' larvae form *within* 'parents' by a process of mitosis from germ cells. The latter are derived from a blastomere separated off at an early stage in the development of the initial sexually-formed zygote.

Asexual reproduction does not occur in the Mollusca and Arthropoda so that it seems to be absent from the more highly organised invertebrates. It is also absent from Nemathelminthes, Rotifera and Tardigrada which exhibit extreme cell constancy and little or no powers of regeneration.

The relative merits of asexual reproduction and gamete formation are difficult to assess because the products of each kind of reproduction are so different. However, one point of comparison is that the tissues

Figure 6.11: Life-cycle of a liver fluke.

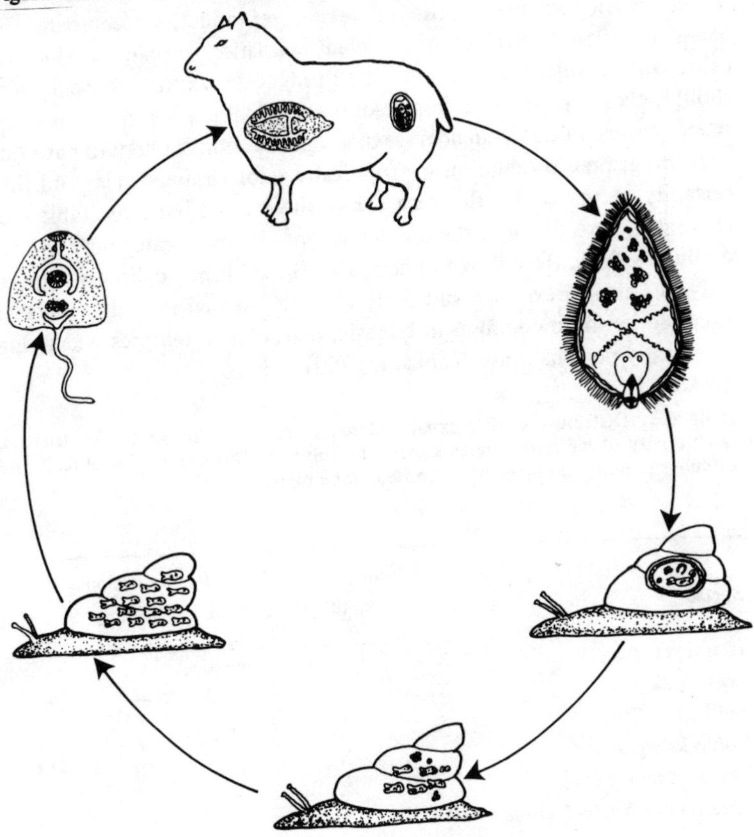

Source: After Cohen (1977). Illustration by L.J. Calow.

set aside for reproduction by the asexual process depend on the normal methods of tissue formation and this is often slower and less efficient than the conversion of food to reproductive tissue during the process of gametogenesis. For example, in sexual triclads the conversion efficiency of food to gametes is between 50 and 70 per cent whereas the conversion efficiency of food to somatic tissue in adults may be less than 10 per cent. The reason for this difference is that in the growth process of most animals, weight-specific food intake and respiratory output converge as size increases (Figure 5.1). Alternatively, small gametes may be formed more quickly and possibly more easily than differentiated, somatic cells so that more may be produced per unit basal metabolism.

However, the products of asexual reproduction are usually larger and better developed than those of sexual reproduction; compare for example a fission product of a triclad, consisting of many millions of cells, with a unicellular zygote. The products of asexual reproduction should, therefore, require less resources and take less time to develop than zygotes. In consequence, asexual reproduction is likely to have the edge on gamete production in trophically poor circumstances and this certainly seems to be the case for freshwater triclads. As Table 6.3 illustrates, here fission is the dominant reproductive strategy in the less-productive lotic (free-flowing) habitats. These ideas are discussed more fully in Calow, Beveridge and Sibly (1979). Asexual reproduction also seems to be more common in benthic, marine invertebrates occupying trophically poor habitats (Thorson, 1950).

Table 6.3: Distribution of asexual and sexual reproduction in British triclads. Productivity of habitats increases from the top to the bottom of the table. X and + denote presence; number of + denote commonness.

Species	Habitat		Reproduction		
	flowing	still	asexual only	both	sexual only
Phagocata vitta	X		+++	+	+
Crenobia alpina	X		+++	++	+
Polycelis felina	X		+++		
Polycelis nigra		X			+++
Polycelis tenuis		X			+++
Dugesia lugubris/polychroa		X			+++
Dendrocoelum lacteum		X			+++

Source: Data from Reynoldson (1961).

6.7.3 *Asexual Reproduction and Rejuvenation*

If there is any ageing in the tissues of the parent this must not be transmitted to asexually-produced offspring or, at least, if it is, the old tissues must be rejuvenated. In triclads and rhabdocoels this appears to be achieved by mitotic replacement of old tissues with new from a stock of embryonic stem cells – the neoblasts – as described in Section 5.10. The latter are protected from ageing processes in the adult because they are held in a metabolically quiescent state. They are distinguished morphologically, for example, by having a nucleus surrounded by only a small

Figure 6.12: Rates of division in repeated head (h) and tail (t) fragments of the rhabdocoel platyhelminth, *Stenostomum incaudatum.*

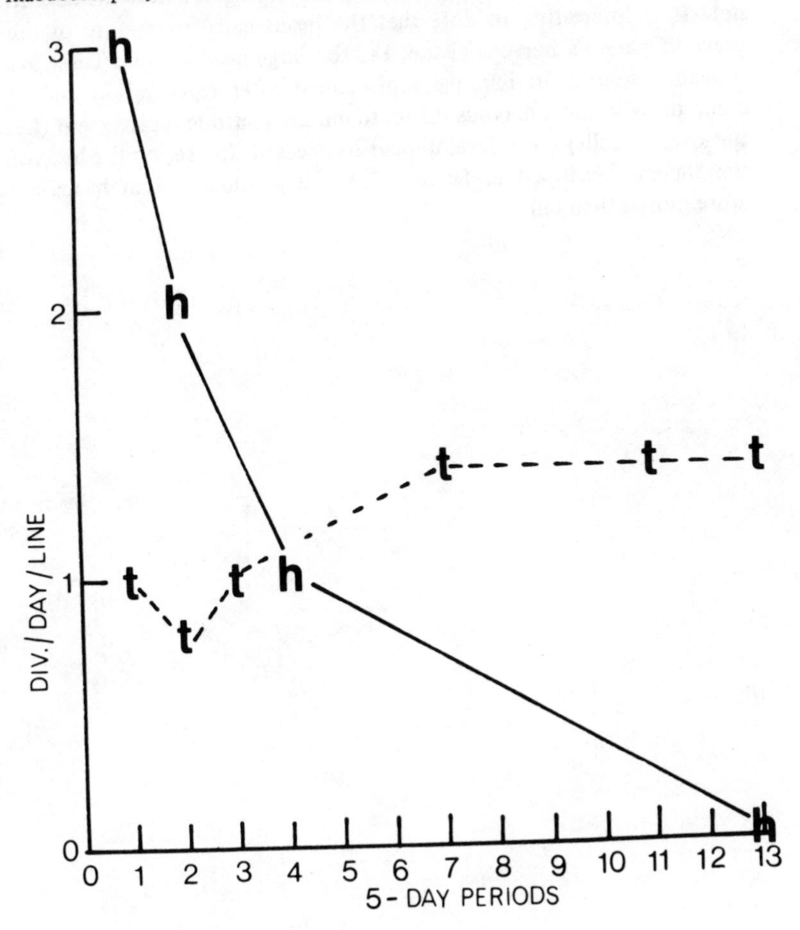

Source: Data from Sonneborn (1930).

amount of cytoplasm and the latter suggests that these cells are not involved in any major metabolic activity.

Of course the transfer of aged tissue and its subsequent replacement may not occur to the same extent in all fission products. Sonneborn (1930), for example, carried out some, now classical, experiments on the rhabdocoel *Stenostomum incaudatum.* This animal reproduces by simple, binary fission. Sonneborn measured rates of fission in fission products derived repeatedly from heads and tails. Some of his results are summarised in Figure 6.12. They show that the rate of reproduction slows down and ultimately stops in the head- but not in the tail-line. The implication is that the heads progressively age whereas the tails do not. It is interesting to note that the heads carry over more of the previous parent's nervous tissue, i.e. the large head ganglia. These are probably subject to minimal replacement after regeneration and in common with most nervous tissue, to minimal mitotic replacement (i.e. turnover of cells) once development has ceased. Hence, on the basis of the theory developed in Section 5.9 it is predictable that heads are more mortal than tails.

7 INTEGRATION

So far we have followed a kind of reductionist methodology – considering the separate elements of the functional biology of invertebrates in complete or, at least, partial isolation. However, organisms interact with their abiotic environment and each other as whole, functional units and so in this final chapter we consider how the separate pieces of the metabolic jig-saw fit together.

7.1 Why the Holistic Approach is Important

It can be argued that a holistic approach is important both in terms of understanding how animals work and how they evolve.

From the physiological point of view animals have to be considered as whole systems because the operation of one part influences that of a constellation of others. For example, we have seen that energy used to power activity is not available for growth and *vice versa,* that that used in the synthesis of storage products is not immediately available for reproduction, that investment in reproduction might influence the repair capacities of the soma and so on. Similarly, a change in the size of an animal, through growth or degrowth, will influence the intensity of the metabolism as a whole, and a change in the relative growth rate of one organ may have profound effects on the form and function of the whole system.

From the evolutionary point of view it is the operation of the whole integrated phenotype which influences the survivorship and fecundity of the organism and hence its relative contribution of offspring to future generations. For example, and as has already been discussed, more resources allocated to reproduction means that more offspring are produced, but with a finite supply of resources this also means that less resources are available for repair processes and active metabolism, both of which might be important for the survival of the parent. Hence the consequences of a particular allocation strategy might have benefits at one level and costs at another such that only a best compromise is likely to have evolved.

The phenotype might be a mosaic of characters and processes controlled by particular genes but its success in the real-world, as measured by the extent it can replicate and pass on genes to future generations, depends upon how well it works out these compromises under the environmental circumstances in which it operates. It is likely, therefore, that patterns of integration will have been subject to selection and that optimal ones (relative to gene transmission) will have evolved.

In what follows, we shall investigate metabolic integration from this adaptational point of view. For more detailed information on the holistic, systems approach to animal development and physiology see Calow (1976) and for an explanation of the application of optimality theory to resource allocation in organisms see Townsend and Calow (1981).

7.2 The Energy Budget as an Integrating Equation

As repeatedly emphasised in the appropriate chapters, growth and reproduction are products of metabolism. They are integrals of energy acquisition and dissipation processes (Figure 5.1) and the energy budget equation provides a way of expressing this, viz:

$$E = \text{input} - \text{losses}$$

where E is the energy available for growth and reproduction. We have aleady seen in Chapter 5 that Winberg restated this equation as:

$$\Delta W/\Delta t = A - R$$

where $\Delta W/\Delta t$ = change in body size, expressed in joules, with time, A = joules of food absorbed across the gut wall from the food and R = joules of heat energy lost as a result of respiratory processes. Warren and Davies (1967) referred to $\Delta W/\Delta t$ as the scope for growth. Hence scope for growth and energy surplus for production amount to one and the same thing.

E is not only a good index of whole-organism physiology, it is also well-correlated with fitness. This is the basis of the maximisation principle, formulated in Chapter 1, i.e. maximisation of E will maximise growth rates, minimise generation times and maximise reproduction – all of which are correlated positively with fitness. We saw in Section 5.5 that this correlation is not perfect, but it has been useful to use the idea that it is, as a first-stage hypothesis in investigating particular physiological strategies from an adaptational point of view.

7.3 Scope for Growth in *Mytilus*

A large number of studies are reported in the literature on the bioenergetics of invertebrates (for a review see Humphreys, 1979), but few are complete or careful enough to give much indication of how investment priorities change through the life-cycles of individuals and under different environmental conditions. An exception is the work of Bayne and co-workers on the mussel, *Mytilus edulis* (summarised and reviewed in Bayne *et al.*, 1976a, b, c).

Bayne and his co-workers report experiments on the effects of temperature and ration on the scope for growth of *Mytilus*. When subjected to a rise in temperature the scope for growth was reduced and when ration level was low the scope became negative (i.e. degrowth occurred). After this, there followed a general acclimation of respiration and feeding rates resulting in the establishment of a new and higher scope for growth. As expected from the theoretical speculation outlined previously, acclimation, here, seems to be a co-ordinated adjustment of key physiological processes to maximise the accumulation of biomass.

Following acclimation, the scope for growth of *Mytilus* remains relatively independent of temperature from 10 to 20°C, although it is markedly dependent on ration (Figure 7.1). Above 20°C there is a decline in the scope for growth. These temperatures approach the lethal limit of the species and here denaturation effects probably dominate over the rate-effects of temperature.

Figure 7.2 examines the effects of ration level and mussel size in more detail. At any given ration, smaller mussels have a greater scope for growth than larger ones. Furthermore, an analysis of the input-output relationships indicates that these smaller and younger animals are more efficient converters of food to tissue than the older ones. Such relationships possibly reflect differences between the weight-exponents of feeding and respiration, for as mussels grow in size the increase in respiratory rate per unit size is disproportionate to the increase in ingestion rate (as depicted in Figure 5.1). The conversion efficiencies of *Mytilus* were also reducing functions of ration up to an optimum, suggesting that over this range mussels became more efficient as rations were reduced. In appropriate chapters we have already seen how such economies might be made.

Using the kinds of ideas introduced here relative to a specific example, we now consider the influences of the two main environmental variables, ration and temperature, on whole-organism metabolism in more general terms.

Figure 7.1: Scope for growth of *Mytilus edulis* (in cal. g^{-1}. hr^{-1}) held at three ration levels: 0.91% (▲), 1.52% (■) and 3.04% (O) of the body weight per day.

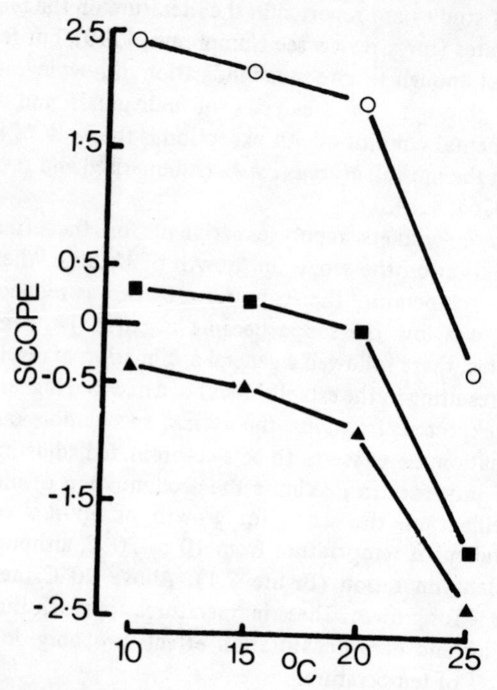

Source: Redrawn from Bayne *et al.* (1976c).

7.4 Integration Under Temperature Stress

Studies in which both the rates of energy input and output have been measured simultaneously are rare for invertebrates. Most information is available on filter-feeders and here the ratio of clearance rate (V_W = feeding rate = litres of water filtered for particles) to oxygen consumption (ml oxygen consumed = V_{O_2}) can be used as an index of energy balance. Since V_W and V_{O_2} are partly independent, various possibilities exist for the adjustment of V_W/V_{O_2} to a temperature disturbance and these have been reviewed by Newell and Branch (1980), viz:

Type 1 Response. Both V_W and V_{O_2} are adjusted; V_W such that maximal rates coincide with environmental temperature and V_{O_2} such that positive acclimation occurs after a change in environmental temperature.

Figure 7.2: Scope for growth of *Mytilus edulis* (% total body calories per day) against ration (% body weight per day) for individuals of different size at 15°C.

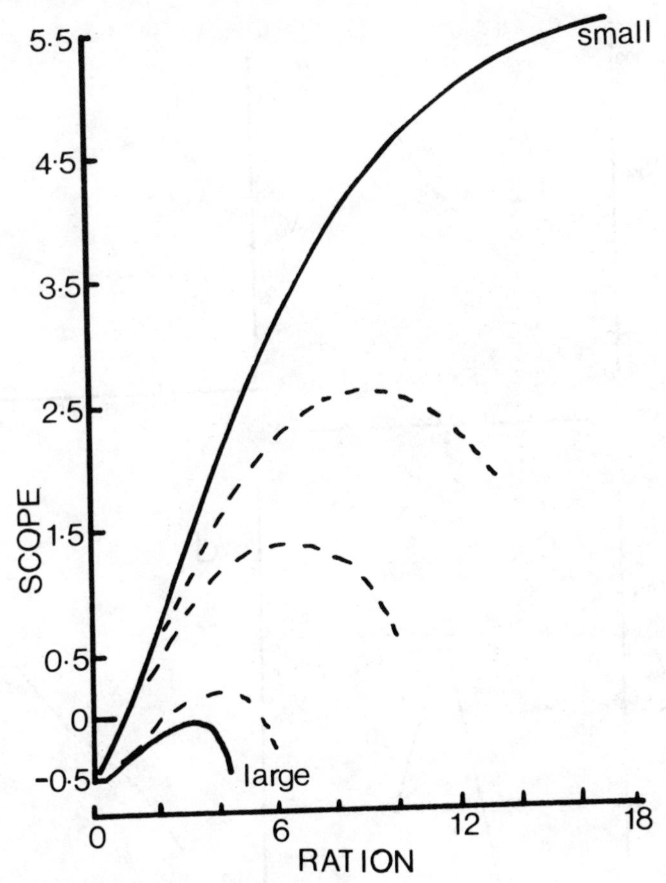

Source: Redrawn from Bayne *et al.* (1976c).

The filter-feeding limpet, *Crepidula fornicata*, shows this type of response and Figures 7.3a and b illustrate respectively how once acclimated, V_W increases with temperature but V_{O_2} is kept constant. Hence V_W/V_{O_2} increases with temperature in this species up to approximately 20°C when V_W/V_{O_2} becomes independent of temperature.

Type 2 Response. Here only V_W is adjusted, and this alone has to compensate for temperature disturbance. The filter-feeding bivalve, *Ostrea edulis,* has this kind of strategy, and the relative changes in acclimated

Figure 7.3: Acclimated values of V_W and V_{O_2} against temperature in *Crepidula fornicata* (a, b) and *Ostrea edulis* (c, d). See text for further explanation.

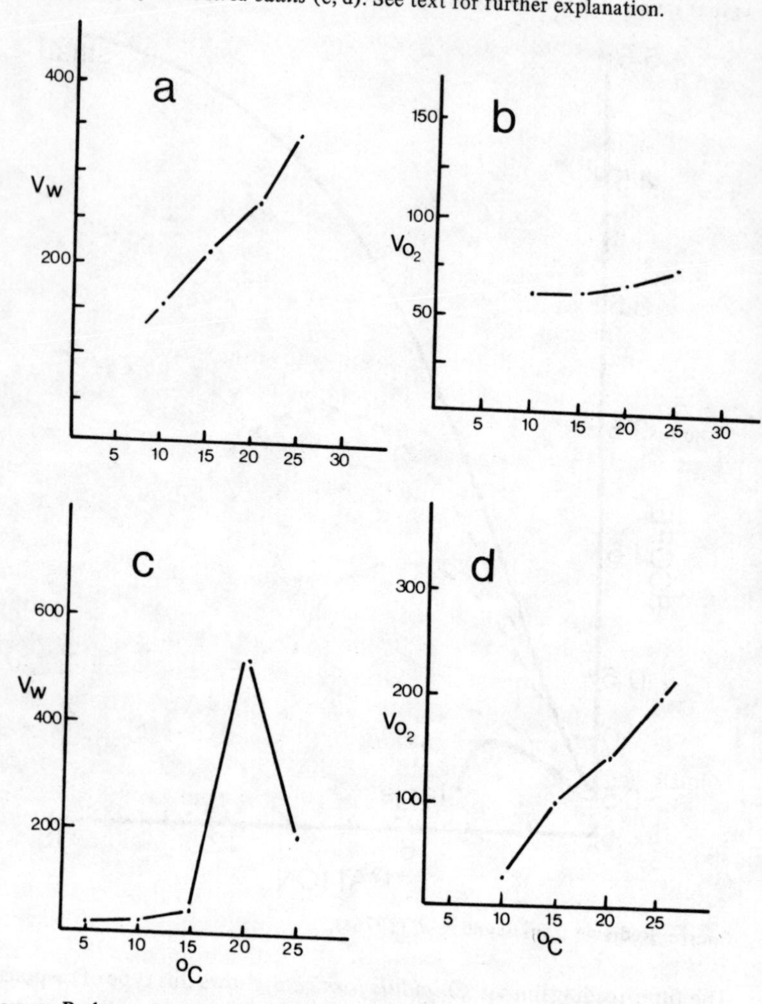

Source: Redrawn from Newell and Branch (1980).

rates to temperature are illustrated in Figure 7.3c and d. In this instance V_W/V_{O_2} increases to approximately 20°C and then declines.

Type 3 Response. Here the energy input is inflexible but V_{O_2} shows compensatory, positive acclimation. The best example of this type of response is from a grazing, not a filter-feeding, snail – *Littorina littorea.*

The rate of radular beat of this species varies with short-term temperature exposure and with the period it has been exposed by the tide. There is some compensation in the R/T curve for radular action – but this is only partial. V_{O_2}, nevertheless, shows more complete acclimation so that the ratio of radular movement to V_{O_2} increases with temperature to 15°C and then declines.

7.5 Integration Under Food Stress

This topic has already been considered in Chapters 2, 5 and 6. Growth does show regulation with respect to feeding disturbances in several invertebrates and this can be traced to modifications in feeding behaviour and respiratory metabolism. It has to be noted, though, that these two variables are not independent. Food-finding and eating are active processes which incur metabolic costs. It has already been stressed that whether a seek-out or sit-and-wait strategy is favoured in the absence of food depends on the type of food in question and on the metabolic state of the feeder.

In general, it is to be anticipated, on the maximisation principle, that respiratory rates should always be minimised. However, this carries other costs. High *rates* of metabolism (ATP generation) are required to sustain high *rates* of conversion and also high levels of activity – both of which can bring positive fitness benefits. Given these opposing costs and benefits it is to be anticipated that respiratory rates are lowest in animals living in poor trophic conditions. For example, intertidal barnacles, by virtue of their filtering habits, have progressively less time to feed the higher they occur in the intertidal zone. Subtidal species have been found to have high metabolic rates, low- and mid-shore species intermediate ones and extremely high-shore species have the lowest rates of all (Newell and Branch, 1980). A similar comparison can also be made for littoral Anthozoa and patellid limpets (Newell and Branch, 1980). Spiders, which are top carnivores and are subject to the continuous possibility of food shortage, have lower metabolic rates than most other similar-sized invertebrates and Anderson (1970) has suggested that this is an energy conservation strategy.

7.6 Modelling Metabolism

The energy budget as stated in Section 7.2 is a reductionistic equation,

Figure 7.4: An integrated model of the bioenergetics of *Mytilus edulis.* The model is based on the separate, in-depth analysis of component processes and was put together for computer simulation.

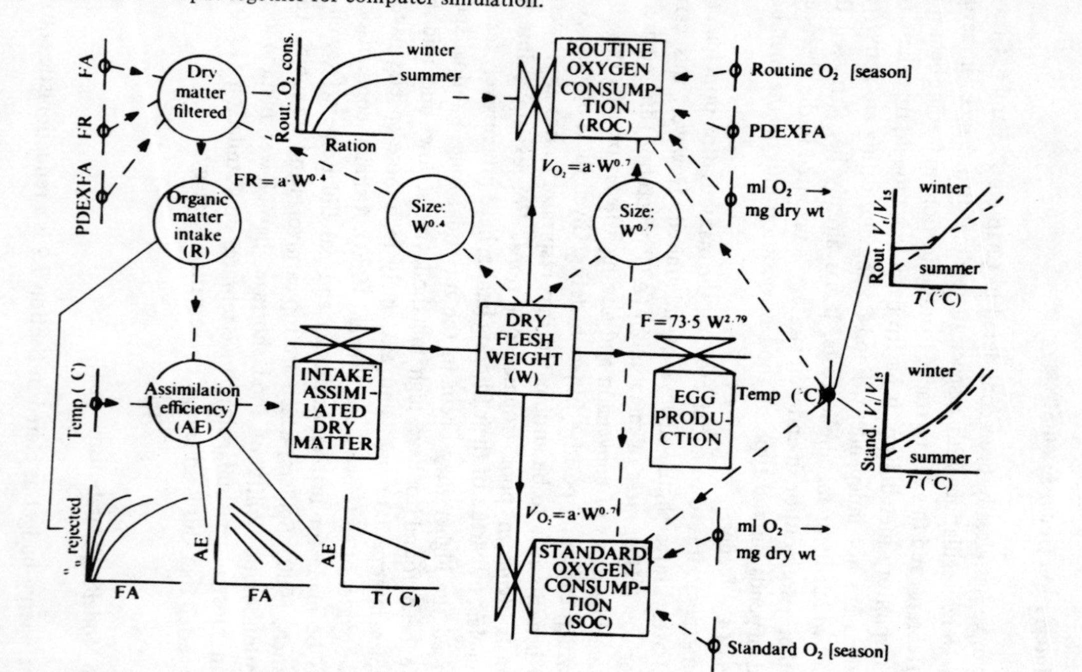

Source: With permission from Bayne *et al.* (1976c).

breaking the metabolic processes down into separate functional parts. One way of trying to add in integration is to study interactions between component processes under what are considered to be ecologically relevant conditions. This is a major task and has only been attempted for one invertebrate – *Mytilus edulis* – by Bayne and co-workers (1976a, b, c). The sort of model that this leads to is shown in Figure 7.4. Another approach is to study the whole-systems properties, such as growth and reproduction, and to use information from these to deduce how the component processes are constrained and controlled to produce the 'desired effect'. This results in the kind of systems-model which is illustrated in Figure 5.8.

REFERENCES

Chapter 1

Alexander, R.McN. (1979). *The Invertebrates.* Cambridge University Press; Cambridge

Barrington, E.J.W. (1967). *Invertebrate Structure and Function.* 1st edn. Nelson; London (see also 2nd edn, 1979)

Calow, P. and Townsend, C.R. (1981). Energetics, ecology and evolution. In: Townsend, C.R. and Calow, P. (eds.) *Physiological Ecology: An Evolutionary Approach to Resource Use.* Blackwells; Oxford

Fretter, V. and Graham, A. (1976). *A Functional Anatomy of Invertebrates.* Academic Press; London, New York and San Francisco

Russell-Hunter, W.D. (1979). *A Life of Invertebrates.* Collier Macmillan Publishers; London

Chapter 2

Baker, J.E. (1978). Midgut clearance and digestive enzyme levels in larvae of *Attagenus megatoma* following removal from food. *Journal of Insect Physiology, 24,* 133-6

Barton Browne, L. and Evans, D.R. (1960). Locomotor activity of blowfly as a function of feeding and starvation. *Journal of Insect Physiology, 4,* 27-37

Bolton, P.J. and Phillipson, J. (1976). Burrowing, feeding, egestion and energy budgets of *Allolobophora rosea* (Savigny) (Lumbricidae). *Oecologia, 23,* 225-45

Boucher-Rodoni, R. (1973). Vitesse de digestion d'*Octopus cyanea* (Cephalopoda: Octopoda). *Marine Biology, 18,* 237-42

Buhr, K-J. (1976). Suspension-feeding and assimilation efficiency in *Lanice conchilega* (Polychaeta). *Marine Biology, 38,* 373-83

Calow, P. (1974a). Evidence for bacterial feeding in *Planorbis contortus* Linn. (Gastropoda: Pulmonata). *Proceedings of the Malacological Society of London, 41,* 145-56

—— (1974b). Some observations on locomotory strategies and their metabolic effects in two species of freshwater gastropods, *Ancylus fluviatilis* Müll. and *Planorbis contortus* Linn. *Oecologia, 16,* 149-61

—— (1975). The feeding strategies of two freshwater gastropods,

Ancylus fluviatilis Müll. and *Planorbis contortus* Linn. (Pulmonata), in terms of ingestion rates and absorption efficiencies. *Oecologia, 20,* 33-49

—— (1977). Ecology, evolution and energetics: a study in metabolic adaptation. *Advances in Ecological Research, 10,* 1-62

—— (1979). Why some metazoan mucus secretions are more susceptible to microbial attack than others. *American Naturalist, 114,* 149-52

—— (1980). Feeding in freshwater triclads – adaptational aspects. In: Smith, D.C. and Tiffon, Y. (eds.) *Nutrition in Lower Metazoa.* Pergamon Press; London

Calow, P. and Calow, L.J. (1975). Cellulase activity and niche separation in freshwater gastropods. *Nature* (London), *255,* 478-80

Calow, P., Davidson, A.F. and Woollhead, A.S. (1981). Life-cycle and feeding strategies of freshwater triclads: a synthesis. *Journal of Zoology, 193,* 215-37

Calow, P. and Woollhead, A.S. (1977). Locomotory strategies and their effects on the energetics of degrowth of freshwater triclads. *Oecologia, 27,* 353-67

Carefoot, T.H. (1973). Feeding, food preference and the uptake of food energy by the supralittoral isopod *Ligia pallasii. Marine Biology, 18,* 228-36

Connolly, K.J. (1966). Locomotory activity in *Drosophila* as a function of food deprivation. *Nature* (London), *209,* 224

Cook, R.M. and Cockrell, B.J. (1978). Predator ingestion rate and its bearing on feeding time and the theory of optimal diets. *Journal of Animal Ecology, 47,* 529-47

Cummins, K.W. and Wuychek, J.C. (1971). Caloric equivalents for investigation in ecological energetics. *Communications of the International Association of Theoretical and Applied Limnology, 18,* 1-158

Dethier, V.G. (1976). *The Hungry Fly.* Harvard University Press; Cambridge, Mass.

Doyle, R.W. (1979). Ingestion rate of a selective deposit feeder in a complex mixture of particles: testing the energy-optimization hypothesis. *Limnology and Oceanography, 24,* 867-74

Elner, R.W. and Hughes, R.N. (1978). Energy maximization in the diet of *Carcinus maenas. Journal of Animal Ecology, 47,* 103-16

Engelmann, M.D. (1966). Energetics, terrestrial field studies and animal productivity. *Advances in Ecological Research, 3,* 73-115

Ford, M.J. (1977). Energy costs of the predatory strategy of the web-spinning spider *Lepthyphantes zimmermanni* Bertkau (Linyphiidae). *Oecologia, 28,* 341-9

Frost, B.W. (1977). Feeding behaviour of *Calanus pacificus* in mixtures of food particles. *Limnology and Oceanography, 22*, 761-89

Gelperin, A. (1971). Regulation of feeding. *Annual Review of Entomology, 16*, 365-73

Griffiths, D. (1975). Prey availability and the food of predators. *Ecology, 56*, 1209-14

—— (1980). The feeding biology of ant-lion larvae: prey-capture, handling and utilization. *Journal of Animal Ecology, 49*, 99-125

Hargrave, B.T. (1970). The utilization of benthic microflora by *Hyalella azteca* (Amphipoda). *Journal of Animal Ecology, 39*, 427-37

Hassall, M. and Jennings, J.B. (1975). Adaptive features of gut structure and digestive physiology in the terrestrial isopod *Philoscia muscorum* (Scopoli) 1763. *Biological Bulletin, 149*, 348-64

Jørgensen, C.B. (1976). August Pütter, August Krogh and modern ideas on the use of dissolved organic matter in aquatic environments. *Biological Reviews, 51*, 291-329

Krebs, J.R. (1978). Optimal foraging: decision rules for predators. In: Krebs, J.R. and Davies, N.B. (eds.) *Behavioural Ecology*. Blackwells; Oxford

Lawton, J.H. (1970). Feeding and food energy assimilation in larvae of the damsel-fly, *Pyrrhosoma nymphula* (Sutz.) (Odonata: Zygoptera). *Journal of Animal Ecology, 39*, 669-84

Lawton, J.H., Beddington, J.R. and Bonser, R. (1974). Switching in invertebrate predators. In: Usher, M.B. and Williamson, M.H. (eds.) *Ecological Stability*. Chapman and Hall; London

Lehman, J.T. (1976). The filter-feeder as an optimal forager and the predicted shapes of feeding curves. *Limnology and Oceanography, 21*, 501-16

Maiorana, V.C. (1979). Nontoxic toxins: the energetics of coevolution. *Biological Journal of the Linnean Society, II*, 387-96

Mansour-Bek, J.J. (1954). The digestive enzymes in Invertebrata and Protochordata. *Tabulae Biologicae, 21*, part 3, no. 24, 75-382

McCleery, R.H. (1977). On satiation curves. *Animal Behaviour, 25*, 1005-15

Menge, B. (1972). Foraging strategies in starfish in relation to actual prey availability and environmental predictability. *Ecological Monographs, 42*, 25-50

Murdoch, W.W. (1969). Switching in general predators: experiments on predator specificity and stability of prey populations. *Ecological Monographs, 39*, 335-54

Phillipson, J. (1964). A miniature bomb calorimeter for small biological samples. *Oikos, 15*, 130-9

Reeve, M.R. (1963). The filter-feeding of *Artemia* I in pure cultures of plant cells. *Journal of Experimental Biology, 40*, 195-205

Reynierse, J.H., Manning, A. and Cafferty, D. (1972). Effects of hunger and thirst on body weight and activity in the cockroach *Nauphoeta cinerea. Animal Behaviour, 20*, 751-7

Richardson, A.M.M. (1975). Food, feeding rates and assimilation in the land snail, *Cepaea nemoralis* L. *Oecologia, 19*, 59-70

Schoener, T.W. (1971). Theory of feeding strategies. *Annual Review of Ecology and Systematics, 2*, 369-404

Sibly, R. (1981). Strategies of digestion, absorption and defecation. In: Townsend, C.R. and Calow, P. (eds.) *Physiological Ecology; An Evolutionary Approach to Resource Use.* Blackwells; Oxford

Southwood, T.R.E. (1961). The number of species of insects associated with various trees. *Journal of Animal Ecology, 30*, 1-8

—— (1973). The insect/plant relationship – an evolutionary perspective. In: van Embden, H.F. (ed.) *Insect/Plant Relationships. Symposium of the Royal Entomological Society, 6.* Blackwells; Oxford

Taghon, G.L., Self, P.A. and Jumars, P.A. (1978). Predicting particle selection by deposit feeders: a model and its implications. *Limnology and Oceanography, 23*, 752-9

Vadas, R.L. (1977). Preferential feeding; an optimization strategy in sea urchins. *Ecological Monographs, 47*, 337-71

Waldbauer, C.P. (1968). The consumption and utilization of food by insects. *Advances of Insect Physiology, 5*, 282-8

Wilson, D.S. (1975). The adequacy of body size as a niche difference. *American Naturalist, 109*, 769-84

Yonge, C.M. (1928). Feeding mechanisms in the invertebrates. *Biological Reviews, 3*, 21-76

—— (1937). Evolution and adaptation in the digestive system of the metazoa. *Biological Reviews, 12*, 87-115

Chapter 3

Alderice, D.F. (1972). Factor combinations. In: Kinne, O. (ed.) *Marine Ecology*, vol. I (3). Wiley Interscience; London

Bayne, B.L. and Scullard, C. (1977). An apparent specific dynamic action in *Mytilus edulis* L. *Journal of the Marine Biological Association of the UK, 57*, 371-8

Bayne, B.L., Thompson, R.J. and Widdows, J. (1976). Physiology 1. In:

Bayne, B.L. (ed.) *Marine Mussels: Their Ecology and Physiology*. Cambridge University Press; Cambridge

Brody, S. (1945). *Bioenergetics and Growth*. Hafner; New York

Calow, P. (1975). The respiratory strategies of two species of freshwater gastropods (*Ancylus fluviatilis* Müller and *Planorbis contortus* Linn.) relative to temperature, oxygen concentration, body size and season. *Physiological Zoology, 48*, 114-29

Conover, R.J. (1956). Oceanography of Long Island Sound 1952-5. VI Biology of *Arcatia clausi* and *A. tonsa*. *Bulletin of the Bingham Oceanography College, 15*, 156-233

Dame, R.F. and Vernberg, F.J. (1978). The influence of constant and cyclic acclimation temperatures on the metabolic rates of *Panopeus herbstii* and *Uca pugilator*. *Biological Bulletin, 154*, 188-97

Davies, P.S. (1966). Physiological ecology of *Patella* I. The effect of body size and temperature on metabolic rate. *Journal of the Marine Biological Association of the U.K., 46*, 647-58

Davies, P.S. and Tribe, N.A. (1969). Temperature dependence of metabolic rate in animals. *Nature* (London), *224*, 723-4

Dejours, P. (1981). *Principles of Comparative Respiratory Physiology*. 2nd edn Elsevier/North Holland Publishing Co.; Amsterdam

Denny, M. (1980) Locomotion: the cost of gastropod crawling. *Science, 208*, 1288-9

Ellenby, C. (1953). Oxygen consumption and cell size. A comparison of the rate of oxygen consumption of diploid and triploid prepupae of *Drosophila melanogaster* Meigen. *Journal of Experimental Biology, 30*, 475-91

Fry, F.E.J. (1958). Temperature compensation. *Annual Review of Physiology, 20*, 207-24

Halcrow, K. and Boyd, C.M. (1967). The oxygen consumption and swimming activity of the amphipod *Gammarus oceanicus* at different temperatures. *Comparative Biochemistry and Physiology, 23*, 233-42

Hemmingsen, A.M. (1960). Energy metabolism as related to body-size and respiratory surfaces and its evolution. *Reports of the Steno Memorial Hospital and the Nordisk Insulinlaboratorium, 9*, 1-110

Hochachka, P.W. (1976). Design of metabolic and enzyme machinery to fit lifestyle and environment. *Biochemical Society Symposium, 41*, 3-31

—— (1980). *Living Without Oxygen; Closed and Open Systems in Hypoxia Tolerance*. Harvard University Press; Cambridge, Mass.

Hochachka, P.W. and Somero, G.N. (1973). *Strategies of Biochemical Adaptation*. W.B. Saunders; Philadelphia

Hughes, C.M. and Shelton, G. (1962). Respiratory mechanisms and their nervous control in fish. *Advances in Comparative Physiology and Biochemistry, 1*, 275-364

Jones, D.R. (1971). Theoretical analysis of factors which may limit the maximum oxygen uptake of fish. The oxygen cost of the cardiac and branchial pumps. *Journal of Theoretical Biology, 32*, 341-9

Jones, J.D. (1972). *Comparative Physiology of Respiration.* Edward Arnold; London

Krogh, A. (1941). *The Comparative Physiology of Respiratory Mechanisms.* University of Pennsylvania Press; Philadelphia

Krogh, A. and Weis-Fogh, T. (1951). The respiratory exchange of the desert Locust (*Schistocerca gregaria*) before, during and after flight. *Journal of Experimental Biology, 28*, 344-57

Lehninger, A.L. (1973). *Bioenergetics.* W.A. Benjamin; Menlo Park, California

Marshall, S.M., Nichols, A.G. and Orr, A.P. (1934). On the biology of *Calanus finmarchicus.* Part VI. Oxygen consumption in relation to environmental conditions. *Journal of the Marine Biological Association of the U.K., 20*, 1-25

Mason, C.F. (1971). Respiration rates and population metabolism in woodland snails. *Oecologia, 7*, 80-94

Mitchell, H.H. (1962). *Comparative Nutrition of Man and Domestic Animals.* Academic Press; New York

Nelson, S.G., Knight, A.W. and Li, H.W. (1977). The metabolic cost of food utilization and ammonia production by juvenile *Macrobrachium rosenbergii* (Crustacea: Palaemonidae). *Comparative Biochemistry and Physiology, 57A*, 67-72

Newell, R.C. (1970). *Biology of Intertidal Animals.* Logos Press; London

Newell, R.C. and Pye, V.I. (1971). Variations in the relationship between oxygen consumption, body size and summated tissue metabolism in the winkle *Littorina littorea. Journal of the Marine Biological Association of the U.K., 51*, 315-38

Parry, G.D. (1978). Effects of growth and temperature acclimation on metabolic rate in the limpet *Cellana tramoserica* (Gastropoda: Patellidae). *Journal of Animal Ecology, 47*, 351-68

Phillipson, J. (1981). Bioenergetic options and phylogeny. In: Townsend, C.R. and Calow, P. (eds.) *Physiological Ecology: An Evolutionary Approach to Resource Use,* Blackwells; Oxford

Precht, H. (1958). Concepts of temperature adaptation of unchanging reaction systems of cold-blooded animals. In: Prosser, C.L. (ed.) *Physiological Adaptation.* Ronald Press; New York

Prosser, C.L. (1973). *Comparative Animal Physiology*. 3rd edn. W.B. Saunders; Philadelphia
Richman, S. (1958). The transformation of energy by *Daphnia pulex*. *Ecological Monographs, 28*, 273-91
Weis-Fogh, T. (1954). Fat combustion and the metabolic rate of flying locusts (*Schistocerca gregaria* Fenskal). *Philosophical Transactions of the Royal Society, 237*, 1-36
Wigglesworth, V.B. (1974). *Insect Physiology*. Chapman and Hall; London
Zeuthen, E. (1970). Rate of living as related to body size in organisms. *Polskie Archiwum Hydrobiologii, 17*, 21-30
Zwaan, A. de. and Wijsman, T.C.M. (1976). Anaerobic metabolism in Bivalvia. *Comparative Biochemistry and Physiology, 54B*, 313-24

Chapter 4

Badman, D.G. (1971). Nitrogen excretion in two species of pulmonate land snails. *Comparative Biochemistry and Physiology, 38A*, 663-73
Baldwin, E. (1935). The energy sources in ontogenesis. VIII. The respiratory quotient of developing gastropod eggs. *Journal of Experimental Biology, 12*, 27-35
Duerr, F. (1966). Nitrogen excretion in the freshwater pulmonate snail *Lymnaea stagnalis appressa* (Say). *Physiologist, 9*, 172
—— (1967). The uric acid content of several species of prosobranch and pulmonate snails as related to nitrogen excretion. *Comparative Biochemistry and Physiology, 22*, 333-40
Horne, F.R. (1971). Accumulation of urea by a pulmonate snail during aestivation. *Comparative Biochemistry and Physiology, 38A*, 565-70
Needham, J. (1931). *Chemical Embryology*. 3 vols. Cambridge University Press; Cambridge
—— (1938). Contribution of chemical physiology to the problem of reversibility in evolution. *Biological Reviews, 13*, 225-51
Pilgrim, R.L.C. (1954). Waste of carbon and energy in nitrogen excretion. *Nature* (London), *173*, 491-2
Potts, W.T.M. and Parry, G. (1964). *Osmotic and Ionic Regulation in Animals*. Pergamon Press; London

Chapter 5

Bak, R.P.M. (1976). The growth of coral colonies and the importance of crustose coralline algae and burrowing sponges with carbonate accumulation. *Netherland Journal of Sea Research, 10*, 237-85

Barr, T.C. (1968). Cave ecology and the evolution of troglobites. *Evolutionary Biology, 2*, 35-102

Bertalanffy, L.von (1960). Principles and theory of growth. In: Nowinski, W. (ed.) *Fundamental Aspects of Normal and Malignant Growth.* Elsevier; Amsterdam

Bidder, G.P. (1933). Senescence. *British Medical Journal, ii*, 583-5

Calow, P. (1976). *Biological Machines.* Edward Arnold; London

—— (1977a). Conversion efficiencies in heterotrophic organisms. *Biological Reviews, 52*, 385-409

—— (1977b). Irradiation studies on rejuvenation in triclads. *Experimental Gerontology, 12*, 173-9

—— (1978a). Bidder's hypothesis revisited: a solution to some key problems associated with the general molecular theory of ageing. *Gerontology, 24*, 448-58

—— (1978b). *Life Cycles.* Chapman and Hall; London

—— (1981). Growth in lower invertebrates. In: Rechcigl, M. (ed.) *Comparative Animal Nutrition.* Karger; Berlin

Calow, P. and Jennings, J.B. (1977). Optimal strategies for the metabolism of reserve materials in microbes and Metazoa. *Journal of Theoretical Biology, 65*, 601-3

Calow, P. and Townsend, C.R. (1981). Resource Utilisation in growth. In: Townsend, C.R. and Calow, P. (eds.) *Physiological Ecology: An Evolutionary Approach to Resource Use.* Blackwells; Oxford

Calow, P. and Woollhead, A.S. (1977). The relationship between ration, reproductive effort and age-specific mortality in the evolution of life-history strategies – some observations on freshwater triclads. *Journal of Animal Ecology, 46*, 765-81

Case, T.J. (1978). On the evolution and adaptive significance of postnatal growth rates in terrestrial vertebrates. *Quarterly Review of Biology, 53*, 243-82

Dawydoff, C. (1910). Restitution von Kopfstücken die vorder Mundöffnung abgeschnitter waren, bei den Nemertinen (*Lineus lacteus*). *Zoologischer Anzeiger, 36*, 1-6

Huxley, J.S. (1932). *Problems of Relative Growth.* Methuen; London

Kirkwood, T.B.L. (1980). Repair and its evolution: survival versus reproduction. In: Townsend, C.R. and Calow, P. (eds.) *Physiological Ecology; An Evolutionary Approach to Resource Use.* Blackwells; Oxford

Lawlor, L.R. (1979). Molting, growth and reproductive strategies in the

terrestrial isopod, *Armadillidium vulgare*. *Ecology*, *57*, 1179-94
Minot, C.S. (1889). Senescence and rejuvenation. *Journal of Physiology*, *12*, 97-153
Odum, H.T. and Pinkerton, R. (1955). Time's speed regulator. *American Scientist*, *43*, 341-3
Rothstein, M. (1979). The formation of altered enzymes in ageing animals. *Mechanisms of Ageing and Development*, *9*, 197-202
Sebens, K.P. (1979). The energetics of asexual reproduction and colony formation in benthic marine invertebrates. *American Zoologist*, *19*, 685-97
Wigglesworth, V.B. (1974). *Insect Physiology*. Chapman and Hall; London
Winberg, G.G. (1956). Rate of metabolism and food requirements of fishes. *Fisheries Research Board of Canada Translation Series*, *No. 194*, 1-253

Chapter 6

Blower, G. (1969). Age-structure of millipede populations in relation to activity and dispersion. In: Sheal, J.G. (ed.) *The Soil Ecosystem* (Symposium volume) *Systematics Assoc.*, *8*, 209-16
Calow, P. (1979). The cost of reproduction – a physiological approach. *Biological Reviews*, *54*, 23-40
Calow, P., Beveridge, M. and Sibly, R. (1979). Heads and tails; adaptational aspects of asexual reproduction in triclads. *American Zoologist*, *19*, 715-27
Calow, P., Davidson, A.F. and Woollhead, A.S. (1981). Life-cycle and reproductive strategies in triclads – a synthesis. *Journal of Zoology*, *193*, 215-37
Christiansen, F.B. and Fenchel, T.M. (1979). Evolution of marine invertebrate reproductive patterns. *Theoretical Population Biology*, *16*, 267-82
Cohen, J. (1977). *Reproduction*. Butterworths; London
Cole, L.C. (1954). The population consequences of life-history phenomenon. *Quarterly Review of Biology*, *29*, 103-37
Comfort, A. (1957). The duration of life in molluscs. *Proceedings of the Malacological Society of London*, *32*, 219-49
Dixon, A.F.G. (1973). *Biology of Aphids*. Edward Arnold; London
Duncan, C.J. (1960). The genital systems of the freshwater Basommataphora. *Proceedings of the Zoological Society of London*, *135*, 339-56
Glesener, R.T. and Tilman, D. (1978). Sexuality and the components of

environmental uncertainty; clues from geographical parthenogenesis in terrestrial animals. *American Naturalist*, *112*, 659-73

Heath, D.J. (1977). Simultaneous hermaphroditism: cost and benefit. *Journal of Theoretical Biology*, *64*, 363-73

Maiorana, V.C. (1979). Why do adult insects not moult? *Biological Journal of the Linnean Society*, *11*, 253-8

Maynard Smith, J. (1978). *The Evolution of Sex*. Cambridge University Press; Cambridge

O'Dor, R.K. and Wells, M.J. (1978). Reproductive versus somatic growth: hormonal control in *Octopus vulgaris*. *Journal of Experimental Biology*, *77*, 15-31

Olive, P.J.W. and Clark, R.B. (1978). Physiology of reproduction. In: Mill, P.J. (ed.) *Physiology of Annelids*. Academic Press; London

Oya, Y. (1976). The red sea coral, *Stylophora pistillata*. *Nature* (London), *259*, 479-80

Reynoldson, T.B. (1961). Environment and reproduction in freshwater triclads. *Nature* (London), *189*, 329-30

Snell, T.W. and King, C.E. (1977). Life span and fecundity in rotifers: the cost of reproduction. *Evolution*, *31*, 882-90

Southwood, T.R.E. (1976). Bionomic strategies and population parameters. In: May, R.M. (ed.) *Theoretical Ecology*. Blackwells; Oxford

Sonneborn, T.M. (1930). Genetic studies on *Stenostomum incaudatum* n. sp. I. The nature and origin of differences in individuals formed during vegetative reproduction. *Journal of Experimental Biology*, *57*, 57-108

Thorson, G. (1950). Reproduction and larval ecology of marine bottom invertebrates. *Biological Reviews*, *25*, 1-45

Underwood, A.J. (1974). On models for reproductive strategies in marine benthic invertebrates. *American Naturalist*, *108*, 874-8

Vance, R.R. (1973a). On reproductive strategies in marine benthic invertebrates. *American Naturalist*, *107*, 339-52

—— (1973b). More on reproductive strategies in marine benthic invertebrates. *American Naturalist*, *107*, 353-61

Vepsälänen, K. (1978). Wing dimorphism and diapause in *Gerris*, determination and adaptive significance. In: Dingle, H. (ed.) *Evolution of Insect Migration and Diapause*. Springer; New York

Woollhead, A.S. and Calow, P. (1979). Energy-partitioning strategies during egg production in semelparous and iteroparous triclads. *Journal of Animal Ecology*, *48*, 491-519

Chapter 7

Anderson, J.F. (1970). Metabolic rate of spiders. *Comparative Biochemistry and Physiology, 33*, 51-72

Bayne, B.L., Thompson, R.J. and Widdows, J. (1976a). Physiology 1. In: Bayne, B.L. (ed.) *Marine Mussels: Their Ecology and Physiology*. Cambridge University Press; Cambridge

Bayne, B.L., Widdows, J. and Thompson, R.J. (1976b). Physiology II. In: Bayne, B.L. (ed.) *Marine Mussels: Their Ecology and Physiology*. Cambridge University Press; Cambridge

—— (1976c). Physiological integrations. In: Bayne, B.L. (ed.) *Marine Mussels: Their Ecology and Physiology*. Cambridge University Press; Cambridge

Calow, P. (1976). *Biological Machines*. Edward Arnold; London

Humphreys, W.F. (1979). Production and respiration in animal populations. *Journal of Animal Ecology, 48*, 427-53

Newell, R.C. and Branch, G.M. (1980). The influence of temperature on the maintenance of metabolic energy balance in marine invertebrates. *Advances of Marine Biology, 17*, 329-96

Townsend, C.R. and Calow, P. (eds.) (1981). *Physiological Ecology: An Evolutionary Approach to Resource Use*. Blackwells; Oxford

Warren, C.E. and Davis, G.E. (1967). Laboratory studies on the feeding, bioenergetics and growth of fish. In: Gerking, S.D. (ed.) *The Biological Basis of Freshwater Fish Production*. Blackwells; Oxford

GLOSSARY OF SYMBOLS

A = food (usually energy) absorbed
C = energy acquired from food, a general term for either ingestion or absorption
E = energy surplus for growth and/or reproduction
F_g = fraction of oxygen in a mixture of gases, e.g. air
F_l = fractional concentration of oxygen in a liquid
I = food (usually energy) ingested
m = meal size
$\hat{m}$ = average meal size
M = body mass
N = net energy from food = gain − cost
P_{O_2} = partial pressure of oxygen (usually in mm Hg)
R = metabolic rate and energy expended therein
SDA = specific dynamic action
SDE = specific dynamic effect
T = available feeding time
t = time
t_m = meal time
$\hat{t}_m$ = average meal time
t_i = time between meals
$\hat{t}_i$ = average time between meals
V_W = filtration rate (of a filter-feeder)
V_{O_2} = volume of oxygen consumed
W = energy measure of biomass
$W/\Delta t$ = scope for growth

INDEX OF ORGANISMS

Only metazoan invertebrates mentioned in the text are listed; they are grouped into phyla, classes and sometimes subclasses and then listed alphabetically. Reference should also be made to Table 1.1 and Figure 1.1.

SUBJECT INDEX